Computational Techniques of Rotor Dynamics with the Finite Element Method

Computational Techniques of Rotor Dynamics with the Finite Element Method

Arne Vollan
Louis Komzsik

CRC Press
Taylor & Francis Group
Boca Raton   London   New York

CRC Press is an imprint of the
Taylor & Francis Group, an **informa** business

CRC Press
Taylor & Francis Group
6000 Broken Sound Parkway NW, Suite 300
Boca Raton, FL 33487-2742

First issued in paperback 2017

Version Date: 20120217

ISBN 13: 978-1-4398-4770-1 (hbk)
ISBN 13: 978-1-138-07347-0 (pbk)

Library of Congress Cataloging-in-Publication Data

Vollan, Arne.
    Computational techniques of rotor dynamics with the finite element method / Arne Vollan and Louis Komzsik.
      p. cm.
    Summary: "This book covers using practical computational techniques for simulating behavior of rotational structures and then using the results to improve fidelity and performance. Applications of rotor dynamics are associated with important energy industry machinery, such as generators and wind turbines, as well as airplane engines and propellers. This book presents techniques that employ the finite element method for modeling and computation of forces associated with the rotational phenomenon. The authors also discuss state-of-the-art engineering software used for computational simulation, including eigenvalue analysis techniques used to ensure numerical accuracy of the simulations"-- Provided by publisher.
    Includes bibliographical references and index.
    ISBN 978-1-4398-4770-1 (hardback)
    1. Rotors--Dynamics. 2. Finite element method. I. Komzsik, Louis. II. Title.

TJ1058.V65 2012
621.8'2--dc23
                                             2012001338

Visit the Taylor & Francis Web site at
http://www.taylorandfrancis.com

and the CRC Press Web site at
http://www.crcpress.com

Contents

Part II   Engineering Analysis of Rotating Structures

Preface

Rotor dynamics is both a classical and a modern branch of engineering science. The rotation of rigid bodies, mainly those with regular shapes such as cylinders and shafts, has been well understood for more than a century. However, the rotational behavior of flexible bodies, especially those with irregular shapes like propeller and turbine blades, requires more modern tools such as finite elements, hence the title and focus of the book. The book is composed of two parts, the first focusing on the theoretical foundation of rotor dynamics and the second on the engineering analysis of rotating structures.

The rotational phenomenon itself is instrumental in our lives. The effects of the rotational phenomenon range from the well-known centrifugal force, through the Coriolis force, and to the lesser known Euler force. Chapter 1 of the book reviews these concepts in detail to establish a foundation for the discourse. The computation of these forces is a very important aspect of rotor dynamics. Industrial structures contain both stationary and rotating components, and the coupling of these is another important aspect of rotor dynamics and the topic of Chapter 2.

The computational simulation of these effects is most accurately done with the finite element technique, which is the subject of Chapter 3. Rotor dynamic simulations need efficient computational and accurate numerical techniques that will also be explored in detail in Chapters 4 and 5 of the book.

In the second part of the book, Chapter 6 starts with the interpretation of the computational results that is instrumental in establishing the stable operational range of rotating machinery, which is discussed in detail. The dynamic response of rotational structures to external and internal excitations is explained by practical application examples in Chapter 7. The computational complexity of rotor dynamics is illustrated by a finite element case study in Chapter 8. Finally, the analysis of propellers that are important in the transportation industry and turbines that are instrumental in the energy industry is the specific focus of the concluding Chapters 9 and 10.

This book is a self-contained volume supporting practicing engineers in all engineering fields encountering rotational phenomena. It is also a text for a graduate-level audience, but undergraduates studying this area of engineering will also find it useful. Researchers and members of academia teaching various aspects of the topic may also find the book a useful reference.

Acknowledgments

We are very grateful to Paul Sicking of Siemens PLM for carefully evaluating the mathematical foundation and Ralf Baumann of Lucerne University of Technology for diligently reviewing the engineering aspects. Special thanks are also due to Leonard Hoffnung, Ben-Shan Liao, and Wei Zhang of Siemens PLM and Jasper Egge of Aerodyn Energiesystems for detailed proofreading of specific chapters. Their recommendations and corrections greatly improved the clarity and quality of the book.

We thank Daniel Dörig and Patrice Verdan of AeroFEM GmbH for providing several figures for the book and the image of the finite element model that appears on the cover, as well as BARD Holding GmbH–Offshore Windkraftanlagen, Germany, for providing the windmill picture that also appears on the cover.

We would also like to acknowledge the staff at Taylor & Francis Group/CRC Press for their contributions to this work, especially Nora Konopka, publisher, for her enthusiastic support of the project; Kari Budyk, project coordinator, for her prompt assistance; and David Tumarkin, project editor, for his valuable corrections. We also thank John Gandour for the cover art, and the people at Cenveo, Inc. for typesetting.

Finally, we also thank our respective companies, Siemens PLM and AeroFEM GmbH, for the use of some tools and models, specifically NX NASTRAN (Siemens PLM Software) and GAROS (AeroFEM), which were used to execute the analyses.

Arne Vollan and Louis Komzsik

About the Authors

Arne Vollan studied aeronautical engineering at the Technical University of Trondheim (Norway) and Aachen (Germany), and holds the degree Diplom Ingenieur.

He was employed by several aeronautical companies such as VFW-Fokker (now Airbus), Helicopter Technik Muenchen, Dornier, Nationaal Lucht- en Ruimtevaartlaboratorium, and Pilatus Aircraft as a dynamic and aeroelastic specialist. He was also a consultant and developed programs for the analysis of rotating structures like wind turbines and propellers. Since 2002 he has been working at AeroFEM GmbH in Switzerland on rotor dynamics and the aeroelasticity of aircraft and large wind turbines.

Louis Komzsik is a graduate of the Technical University of Budapest with an engineering degree and the Eötvös University of Sciences in Budapest with a mathematics degree, and he holds a Doctorate from the Technical University of Budapest, Hungary.

He was employed by the Hungarian Shipyards from 1972 to 1980 and worked at the McDonnell-Douglas Corporation in 1981 and 1982. He was the chief numerical analyst at the MacNeal-Schwendler (now MSC Software) Corporation for two decades. Since 2003 he has been the chief numerical analyst at Siemens PLM Software. For the past 30 years he has been the architect of the modern numerical methods of NASTRAN, the world's leading finite element analysis tool in structural engineering.

Arne Vollan (right) and Louis Komzsik (left) cooperated on several development projects during the past quarter century, among them the implementation of the rotor dynamics capability into NX NASTRAN. This book is an outcome of that cooperation and opens up the authors' wealth of experience to a wide engineering audience in the important technical area of rotor dynamics.

Part I

Theoretical Foundation of Rotor Dynamics

1

Introduction to Rotational Physics

The focus of this chapter is to lay the foundation for rotor dynamics. We discuss the various coordinate systems involved and the transformations between them. The kinetic energy of a rotating particle will lead to the equations of motion of uncoupled scenarios.

1.1 Fixed Coordinate System

The interpretation of rotational phenomena requires the introduction of a rotating coordinate system in relation to the fixed coordinate system of the environment. Figure 1.1 depicts the relationship between the two coordinate systems. The location of a point in the fixed system defined by coordinate axes $\bar{x}, \bar{y}, \bar{z}$ is specified by the vector

$$\{\bar{r}\} = \{\sigma\} + \{r\},$$

where $\{r\}$ is the location vector of the point with respect to the moving coordinate system whose coordinate axes are defined by the unit vectors $\{u_1\}$ $\{u_2\}$, $\{u_3\}$. In the equations of this book the curly bracket notation is used to denote vectors. The origin of the moving coordinate system is defined by the vector $\{\sigma\}$. The location vector of the point in the fixed system is defined in terms of the unit coordinate vectors of the rotating system as

$$\{\bar{r}\} = \{\sigma\} + r_1\{u_1\} + r_2\{u_2\} + r_3\{u_3\} = \{\sigma\} + \sum_{i=1}^{3} r_i\{u_i\}. \tag{1.1}$$

We will simplify this very general scenario by assuming that the origin of a rotating coordinate system is coincident with the fixed system (i.e., $\{\sigma\} = 0$) in this chapter. We will later relax this constraint. Let us now view the motion of the point and compute its velocity by simply differentiating as

$$\frac{d\{\bar{r}\}}{dt} = \sum_{i=1}^{3} \frac{dr_i}{dt}\{u_i\} + \sum_{i=1}^{3} r_i \frac{d\{u_i\}}{dt}. \tag{1.2}$$

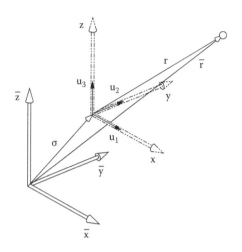

FIGURE 1.1
Coordinate system relations.

The first term on the right-hand side represents the velocity of the point with respect to the rotating coordinate system. The second term represents a (yet somewhat mystic) velocity component related to the temporal change of the unit vectors of the rotating coordinate system. Introducing the corresponding velocity terms to ease the notation, we may write

$$\{\bar{v}\} = \sum_{i=1}^{3} v_i \{u_i\} + \sum_{i=1}^{3} r_i \frac{d\{u_i\}}{dt} = \{v\} + \sum_{i=1}^{3} r_i \frac{d\{u_i\}}{dt}. \tag{1.3}$$

Our interest in the physics of the rotating phenomenon leads us toward the acceleration of the point. Continuing the simple differentiation of the rotating system velocity yields

$$\frac{d\{v\}}{dt} = \sum_{i=1}^{3} \frac{dv_i}{dt} \{u_i\} + \sum_{i=1}^{3} v_i \frac{d\{u_i\}}{dt} = \{a\} + \sum_{i=1}^{3} v_i \frac{d\{u_i\}}{dt}. \tag{1.4}$$

The differentiation of the second term in Equation (1.3) results in

$$\frac{d}{dt} \sum_{i=1}^{3} r_i \frac{d\{u_i\}}{dt} = \sum_{i=1}^{3} v_i \frac{d\{u_i\}}{dt} + \sum_{i=1}^{3} r_i \frac{d^2\{u_i\}}{dt^2}. \tag{1.5}$$

The acceleration of the point with respect to the fixed coordinate system becomes

$$\frac{d^2\{\bar{r}\}}{dt^2} = \{\bar{a}\} = \{a\} + 2\sum_{i=1}^{3} v_i \frac{d\{u_i\}}{dt} + \sum_{i=1}^{3} r_i \frac{d^2\{u_i\}}{dt^2}. \tag{1.6}$$

The first term on the right-hand side is the acceleration with respect to the rotating system and the other terms will be interpreted in the next section.

1.2 Rotating Coordinate System

We now consider the rotating coordinate system. Let the axis of rotation be defined in terms of the unit vectors by the vector

$$\{\Omega\} = \Omega_1\{u_1\} + \Omega_2\{u_2\} + \Omega_3\{u_3\}. \tag{1.7}$$

The components of the rotational vector are the angular velocities with respect to the coordinate axes:

$$\Omega_i = \frac{d\theta_i}{dt}. \tag{1.8}$$

The mysterious first derivative term of the previous section, in the case of the rotating moving coordinate system, is computed as

$$\frac{d\{u_i\}(t)}{dt} = \{\Omega\} \times \{u_i\}(t). \tag{1.9}$$

The second derivative term in the same vein is

$$\frac{d^2\{u_i\}(t)}{dt^2} = \frac{d\{\Omega\}}{dt} \times \{u_i\} + \{\Omega\} \times \frac{d\{u_i\}(t)}{dt} = \frac{d\{\Omega\}}{dt} \times \{u_i\} + \{\Omega\} \times (\{\Omega\} \times \{u_i\}(t)). \tag{1.10}$$

With these formulae, the acceleration of the point developed at the end of the last section becomes

$$\{\bar{a}\} = \{a\} + 2\sum_{i=1}^{3} v_i \{\Omega\} \times \{u_i\}(t) + \sum_{i=1}^{3} r_i \frac{d\{\Omega\}}{dt} \times \{u_i\} + \sum_{i=1}^{3} r_i \{\Omega\} \times (\{\Omega\} \times \{u_i\}(t)). \tag{1.11}$$

Further simplification is possible because the rotation vector is independent of the summation:

$$\{\bar{a}\} = \{a\} + 2\{\Omega\} \times \sum_{i=1}^{3} v_i \{u_i\}(t) + \frac{d\{\Omega\}}{dt} \times \sum_{i=1}^{3} r_i \{u_i\} + \{\Omega\} \times (\{\Omega\} \times \sum_{i=1}^{3} r_i \{u_i\}(t)). \quad (1.12)$$

Finally, introducing the terms defined earlier, the acceleration of the point with respect to the fixed system becomes

$$\{\bar{a}\} = \{a\} + 2\{\Omega\} \times \{v\} + \frac{d\{\Omega\}}{dt} \times \{r\} + \{\Omega\} \times (\{\Omega\} \times \{r\}). \quad (1.13)$$

1.3 Forces in the Rotating System

We now place a particle with mass m into the position of the point we were following. Several forces define the equilibrium of the moving mass particle. We analyze this equilibrium in the rotating coordinate system; hence we express the acceleration of the particle in the rotating system as

$$\{a\} = \{\bar{a}\} - 2\{\Omega\} \times \{v\} - \frac{d\{\Omega\}}{dt} \times \{r\} - \{\Omega\} \times (\{\Omega\} \times \{r\}). \quad (1.14)$$

The forces acting on the particle may be obtained by using Newton's second law, which means a simple multiplication of the acceleration by the mass. This results in several force components. The only active force in the system is the result of the acceleration with respect to the fixed system:

$$\{\bar{F}\} = m \cdot \{\bar{a}\}. \quad (1.15)$$

All the other forces resulting from the acceleration terms on the right-hand side of Equation (1.14) are inertia forces arising in the rotating system. The Coriolis force is the following quantity:

$$\{F_C\} = -2m(\{\Omega\} \times \{v\}). \quad (1.16)$$

This force is rather peculiar and leads to intriguing phenomena in rotating systems as it is influencing the gyroscopic behavior of the rotating component. When the axis of rotation deviates from being perpendicular to the plane of rotation of the particle, which is a common occurrence in industrial applications, this force's magnitude changes. Assuming that the angle between the plane and the axis of rotation is α, the force will have two components. The component acting in the plane of the rotation is the Coriolis effect:

$$\{F_C\} = -2m\left(\{\Omega\}\times\{v\}\right)\sin(\alpha), \tag{1.17}$$

while the component acting perpendicular to the plane of rotation is the Eotvos effect with magnitude

$$\{F_{CE}\} = -2m\left(\{\Omega\}\times\{v\}\right)\cos(\alpha). \tag{1.18}$$

This effect will aid in retaining the plane of rotation when the axis of rotation is not perpendicular to the plane of rotation. When the axis of rotation is perpendicular to the plane of rotation, by the virtue of the sine function, one obtains the full effect of the Coriolis force.

The third acceleration term of Equation (1.14) produces the Euler force:

$$\{F_E\} = -m\frac{d\{\Omega\}}{dt}\times\{r\}. \tag{1.19}$$

The force occurs when there is a nonzero rate of change in the magnitude of the rotation vector. This is an important phenomenon in the operation of rotating machinery during the speed-up or wind-down process.

Finally, the last term of Equation (1.14) produces the familiar centrifugal force:

$$\{F_{Cf}\} = -m\{\Omega\}\times\left(\{\Omega\}\times\{r\}\right). \tag{1.20}$$

The effect of this in rotational structures will be the particles moving outward toward the perimeter of the rotation circle, resulting in a phenomenon called centrifugal softening. These force components are discussed in more detail in the following sections.

1.4 Transformation between Coordinate Systems

Let us restrict the axis of rotation of the rotating coordinate system to the $\{u_3\}$ axis only. In this case the generic rotational vector simplifies to

$$\{\Omega\} = 0\cdot\{u_1\}+0\cdot\{u_2\}+\Omega\cdot\{u_3\}. \tag{1.21}$$

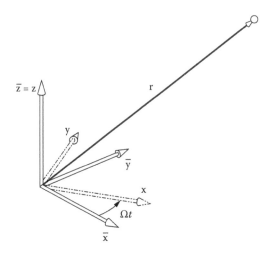

FIGURE 1.2
Transformation between coordinate systems.

Let us further assume that the rotating axis is coincident with one of the axes of the fixed coordinate system, specifically to the z-axis. This results in the form of

$$\{\Omega\} = 0 \cdot \{i\} + 0 \cdot \{j\} + \Omega \cdot \{k\}. \tag{1.22}$$

Finally, we restrict our computations to constant rotational velocity; hence the Euler force will not appear in the formulations. The instantaneous angle between the rotating system and the stationary system is Ωt. This simplified scenario is depicted in Figure 1.2.

The location vector of a particle in the rotating coordinate system is written as the vector

$$r = \left\{ \begin{array}{c} x \\ y \\ z \end{array} \right\}.$$

This location vector is transformed to the fixed system as follows:

$$\bar{r} = \left\{ \begin{array}{c} \bar{x} \\ \bar{y} \\ \bar{z} \end{array} \right\} = \left[\begin{array}{ccc} \cos\Omega t & -\sin\Omega t & 0 \\ \sin\Omega t & \cos\Omega t & 0 \\ 0 & 0 & 1 \end{array} \right] \left\{ \begin{array}{c} x \\ y \\ z \end{array} \right\}.$$

The transformation matrix will be used extensively in the following and it will be denoted

$$[H] = \begin{bmatrix} \cos\Omega t & -\sin\Omega t & 0 \\ \sin\Omega t & \cos\Omega t & 0 \\ 0 & 0 & 1 \end{bmatrix}. \tag{1.23}$$

For the proceeding work we establish some relations of the transformation matrix. The product of the transformation matrix and its transpose produce the identity matrix.

$$[H]^T[H] = \begin{bmatrix} \cos\Omega t & \sin\Omega t & 0 \\ -\sin\Omega t & \cos\Omega t & 0 \\ 0 & 0 & 1 \end{bmatrix} \begin{bmatrix} \cos\Omega t & -\sin\Omega t & 0 \\ \sin\Omega t & \cos\Omega t & 0 \\ 0 & 0 & 1 \end{bmatrix}$$

$$= \begin{bmatrix} 1 & 0 & 0 \\ 0 & 1 & 0 \\ 0 & 0 & 1 \end{bmatrix} = [I]. \tag{1.24}$$

The temporal derivatives of the transformation matrix are also needed and they become

$$[\dot{H}] = \Omega \begin{bmatrix} -\sin\Omega t & -\cos\Omega t & 0 \\ \cos\Omega t & -\sin\Omega t & 0 \\ 0 & 0 & 0 \end{bmatrix} = \Omega[\bar{H}]. \tag{1.25}$$

and

$$[\ddot{H}] = \Omega^2 \begin{bmatrix} -\cos\Omega t & \sin\Omega t & 0 \\ -\sin\Omega t & -\cos\Omega t & 0 \\ 0 & 0 & 0 \end{bmatrix} = \Omega^2[\bar{\bar{H}}].$$

The product of this velocity transformation matrix and the displacement transformation matrix is of the form denoted by P:

$$[\bar{H}]^T[H] = \begin{bmatrix} -\sin\Omega t & \cos\Omega t & 0 \\ -\cos\Omega t & -\sin\Omega t & 0 \\ 0 & 0 & 0 \end{bmatrix} \begin{bmatrix} \cos\Omega t & -\sin\Omega t & 0 \\ \sin\Omega t & \cos\Omega t & 0 \\ 0 & 0 & 1 \end{bmatrix}$$

$$= \begin{bmatrix} 0 & 1 & 0 \\ -1 & 0 & 0 \\ 0 & 0 & 0 \end{bmatrix} = [P]. \tag{1.26}$$

The reverse order product of these matrices results in the same matrix with opposite sign:

$$[H]^T[\bar{H}] = \begin{bmatrix} \cos\Omega t & \sin\Omega t & 0 \\ -\sin\Omega t & \cos\Omega t & 0 \\ 0 & 0 & 1 \end{bmatrix} \begin{bmatrix} -\sin\Omega t & -\cos\Omega t & 0 \\ \cos\Omega t & -\sin\Omega t & 0 \\ 0 & 0 & 0 \end{bmatrix}$$

$$= \begin{bmatrix} 0 & -1 & 0 \\ 1 & 0 & 0 \\ 0 & 0 & 0 \end{bmatrix} = [P]^T = -[P]. \tag{1.27}$$

Finally, the product of the velocity transformation matrix with itself yields another matrix:

$$[\bar{H}]^T[\bar{H}] = \begin{bmatrix} -\sin\Omega t & \cos\Omega t & 0 \\ -\cos\Omega t & -\sin\Omega t & 0 \\ 0 & 0 & 0 \end{bmatrix} \begin{bmatrix} -\sin\Omega t & -\cos\Omega t & 0 \\ \cos\Omega t & -\sin\Omega t & 0 \\ 0 & 0 & 0 \end{bmatrix}$$

$$= \begin{bmatrix} 1 & 0 & 0 \\ 0 & 1 & 0 \\ 0 & 0 & 0 \end{bmatrix} = [J] = [\bar{H}][\bar{H}]^T. \tag{1.28}$$

These relations and matrices will be used extensively in the following sections.

1.5 Kinetic Energy Due to Translational Displacement

The lone particle of our consideration will become a node of a mechanical system; hence we will call its local motion nodal in the following. When the rotating particle undergoes a nodal translation, in the rotating coordinate system it is defined by the vector

$$\{\rho\} = \begin{Bmatrix} u \\ v \\ w \end{Bmatrix}. \tag{1.29}$$

The arrangement is shown in Figure 1.3.

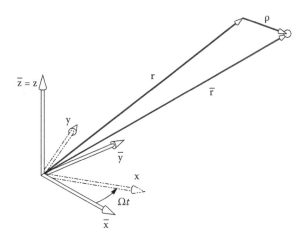

FIGURE 1.3
Translational nodal displacement.

The location vector of the globally rotating and locally displaced particle in the fixed system is

$$\{\bar{r}\} = [H](\{r\} + \{\rho\}). \tag{1.30}$$

Note that the nodal displacement of the particle is also subject to the transformation because it is given with respect to the rotating system. In order to calculate the kinetic energy, the velocity of the particle in the fixed system is needed. Simply differentiating the above vector with respect to time yields

$$\{\dot{\bar{r}}\} = [\dot{H}](\{r\} + \{\rho\}) + [H]\{\dot{\rho}\}, \tag{1.31}$$

where the dot denotes the time derivative. Using the derivative of the transformation matrix introduced in the last section, the velocity vector is expanded as

$$\{\dot{\bar{r}}\} = \Omega[\bar{H}](\{r\} + \{\rho\}) + [H]\{\dot{\rho}\}. \tag{1.32}$$

The kinetic energy by definition is given with respect to the fixed system and is of the form

$$T = \frac{m}{2}\{\dot{\bar{r}}\}^{T}\{\dot{\bar{r}}\}. \tag{1.33}$$

Utilizing the quantities introduced previously yields the kinetic energy in the form

$$T = \frac{m}{2}(\Omega(\{r\}^T + \{\rho\}^T)[\bar{H}]^T + \{\dot{\rho}\}^T[H]^T)(\Omega[\bar{H}](\{r\} + \{\rho\}) + [H]\{\dot{\rho}\}). \qquad (1.34)$$

Employing the terms introduced in the last section and some reordering results in

$$T = \frac{m}{2}(\Omega^2(\{r\}^T + \{\rho\}^T)[J](\{r\} + \{\rho\}))$$

$$+ \frac{m}{2}(\Omega\{\dot{\rho}\}^T[P]^T(\{r\} + \{\rho\}) + \Omega(\{r\}^T + \{\rho\}^T)[P]\{\dot{\rho}\} + \{\dot{\rho}\}^T[I]\{\dot{\rho}\}). \qquad (1.35)$$

By collecting and grouping terms, the kinetic energy for a rotating system analysis becomes

$$T = \frac{m}{2}(\Omega^2\{r\}^T[J]\{r\} + 2\Omega^2\{r\}^T[J]\{\rho\})$$

$$+ \frac{m}{2}(\Omega^2\{\rho\}^T[J]\{\rho\} + 2\Omega\{\dot{\rho}\}^T[P]^T\{r\} + 2\Omega\{\rho\}^T[P]\{\dot{\rho}\} + \{\dot{\rho}\}^T[I]\{\dot{\rho}\}). \qquad (1.36)$$

This will be one component of the equilibrium equation of the rotating particle in a rotating system, considering only the translational displacement's effect in the kinetic energy.

1.6 Kinetic Energy Due to Rotational Displacement

The nodal displacement of the point may also be rotational. We will assume small rotations of the rotating point along the three nodal coordinate axes as

$$\{\alpha\} = \begin{Bmatrix} \varphi \\ \psi \\ \theta \end{Bmatrix}. \qquad (1.37)$$

The scenario is shown in Figure 1.4, where the $A\alpha$ vector now represents the translational displacement of the particle due to the nodal rotation.

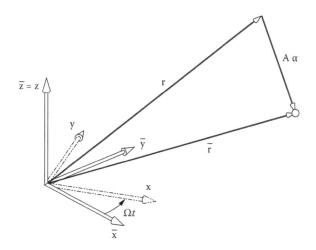

FIGURE 1.4
Rotational nodal displacement.

The nodal rotation requires the consideration of mass and inertia. The center of the mass point is coincident with the node. This is consistent with the lumped mass formulation of the finite element technology. This makes the mass and inertia effects computable without modifying the finite element formulation. This, in turn, provides the additional advantage of utilizing the wide variety of elements available in commercial software.

Inertia moments can be given with the concentrated mass input of commercial finite element codes, or they can be defined in connection with the surrounding mass points in the finite element mesh. It is also possible to define the inertia moments of the mass point with a simple model by attaching six submasses to the node, as shown in Figure 1.5 depicted by the circles.

The total mass at the node is defined by $m = 6m'$. The inertia terms are computed as

$$\Theta_x = m i_x^2 = \frac{m}{3}(y'^2 + z'^2) = 2m'(y'^2 + z'^2),$$

$$\Theta_y = m i_y^2 = \frac{m}{3}(x'^2 + z'^2) = 2m'(x'^2 + z'^2),$$

and

$$\Theta_z = m i_z^2 = \frac{m}{3}(x'^2 + y'^2) = 2m'(x'^2 + y'^2).$$

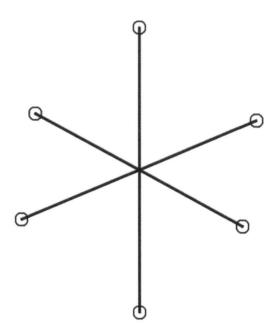

FIGURE 1.5
Node with six masses located at offset x′, y′, and z′ from the node.

Here the terms x', y', z' are the coordinates of the mass points:

$$x'^2 = \frac{3}{2}\left(-i_x^2 + i_y^2 + i_z^2\right),$$

$$y'^2 = \frac{3}{2}\left(i_x^2 - i_y^2 + i_z^2\right),$$

and

$$z'^2 = \frac{3}{2}\left(i_x^2 + i_y^2 - i_z^2\right).$$

Consider first a rotation about the z-axis by angle θ. The transformation of the location of a point in the rotating system due to this rotation is of the form

$$\{\tilde{r}\} = \begin{bmatrix} \cos\theta & -\sin\theta & 0 \\ \sin\theta & \cos\theta & 0 \\ 0 & 0 & 1 \end{bmatrix} \{r'\}. \tag{1.38}$$

A similar transformation for rotation about the x-axis with an angle φ is

$$\{\tilde{r}\} = \begin{bmatrix} 1 & 0 & 0 \\ 0 & \cos\varphi & -\sin\varphi \\ 0 & \sin\varphi & \cos\varphi \end{bmatrix} \{r'\}, \tag{1.39}$$

and one for rotation about the y-axis with an angle ψ is

$$\{\tilde{r}\} = \begin{bmatrix} \cos\psi & 0 & \sin\psi \\ 0 & 1 & 0 \\ -\sin\psi & 0 & \cos\psi \end{bmatrix} \{r'\}. \tag{1.40}$$

The product of the three matrices describes the simultaneous transformation with all three rotations. Taking only the first terms of the Taylor series approximation of sine and cosine results in their replacement by one and angle, respectively. This is adequate for the following derivation except for the centrifugal matrix to be derived in the next section. The accurate derivation of that matrix is shown in the Appendix.

The simplified transformation matrix of a point location due to the simultaneous rotations is written as

$$\{\tilde{r}\} = \begin{bmatrix} 1 & -\theta & 0 \\ \theta & 1 & 0 \\ 0 & 0 & 1 \end{bmatrix} \begin{bmatrix} 1 & 0 & 0 \\ 0 & 1 & -\varphi \\ 0 & \varphi & 1 \end{bmatrix} \begin{bmatrix} 1 & 0 & \psi \\ 0 & 1 & 0 \\ -\psi & 0 & 1 \end{bmatrix}$$

$$\{r\} = \begin{bmatrix} 1 & -\theta & \psi \\ \theta & 1 & -\varphi \\ -\psi & \varphi & 1 \end{bmatrix} \{r'\} = ([I] + [\tilde{A}])\{r'\} \tag{1.41}$$

where

$$[\tilde{A}] = \begin{bmatrix} 0 & -\theta & \psi \\ \theta & 0 & -\varphi \\ -\psi & \varphi & 0 \end{bmatrix}. \tag{1.42}$$

In the matrix the second-order small terms resulting from the matrix multiplication have also been neglected. Let us execute the multiplication:

$$[\tilde{A}]\{r'\} = \begin{bmatrix} 0 & -\theta & \psi \\ \theta & 0 & -\varphi \\ -\psi & \varphi & 0 \end{bmatrix} \begin{Bmatrix} x' \\ y' \\ z' \end{Bmatrix} = \begin{Bmatrix} z'\psi - y'\theta \\ x'\theta - z'\varphi \\ -x'\psi + y'\varphi \end{Bmatrix}. \tag{1.43}$$

Because the scalar multiplications are commutative, this may also be written as

$$\begin{Bmatrix} z'\psi - y'\theta \\ -z'\varphi + x'\theta \\ y'\varphi - x'\psi \end{Bmatrix} = \begin{bmatrix} 0 & z' & -y' \\ -z' & 0 & x' \\ y' & -x' & 0 \end{bmatrix} \begin{Bmatrix} \varphi \\ \psi \\ \theta \end{Bmatrix}. \tag{1.44}$$

Introducing a matrix containing the coordinates of the node's mass points

$$[A] = \begin{bmatrix} 0 & z' & -y' \\ -z' & 0 & x' \\ y' & -x' & 0 \end{bmatrix}, \tag{1.45}$$

the nodal rotations may be transformed into nodal translations by

$$[A]\{\alpha\} = [A] \begin{Bmatrix} \varphi \\ \psi \\ \theta \end{Bmatrix} = \begin{Bmatrix} u \\ v \\ w \end{Bmatrix} = \{\rho\}. \tag{1.46}$$

Here the $\{\rho\}$ represents the translational displacement of the mass particle due to the nodal rotational displacement. With this, the position vector in the fixed system, accounting for the rotational displacement, is of the form

$$\{\bar{r}\} = [H](\{r\} + \{\rho\}) = [H](\{r\} + [A]\{\alpha\}) = [H]\{r\} + [H][A]\{\alpha\}. \tag{1.47}$$

The term on the right-hand side was first adjusted to account for the nodal rotation and then for the global rotation. The velocity of the particle due to the nodal rotational displacement becomes

$$\{\dot{\bar{r}}\} = [\dot{H}]\{r\} + [\dot{H}][A]\{\alpha\} + [H][A]\{\dot{\alpha}\}. \tag{1.48}$$

Using the earlier introduced relations, the velocity is

$$\{\dot{r}\} = \Omega[\bar{H}]\{r\} + \Omega[\bar{H}][A]\{\alpha\} + [H][A]\{\dot{\alpha}\}. \tag{1.49}$$

The kinetic energy computation follows the process executed in the case of the translational displacement and proceeds as

$$T = \frac{1}{2}m(\Omega\{r\}^T[\bar{H}]^T + \Omega\{\alpha\}^T[A]^T[\bar{H}]^T + \{\dot{\alpha}\}^T[A]^T[H]^T)$$

$$+ \frac{1}{2}m(\Omega[\bar{H}]\{r\} + \Omega[\bar{H}][A]\{\alpha\} + [H][A]\{\dot{\alpha}\}). \tag{1.50}$$

Executing the posted multiplications and collecting like terms results in the kinetic energy due to the nodal rotational displacement

$$T = \frac{1}{2}m(\Omega^2\{r\}^T[J]\{r\} + 2\Omega^2\{r\}^T[J][A]\{\alpha\} + 2\Omega\{r\}^T[P][A]\{\dot{\alpha}\})$$

$$+ \frac{1}{2}m(\Omega^2\{\alpha\}^T[A]^T[J][A]\{\alpha\} + 2\Omega\{\alpha\}^T[A]^T[P][A]\{\dot{\alpha}\})$$

$$+ \frac{1}{2}m(\{\dot{\alpha}\}^T[A]^T[I][A]\{\dot{\alpha}\}). \tag{1.51}$$

This will be the basis of further operations to obtain the equation of motion of the particle in the next section.

1.7 Equation of Motion in Rotating Coordinate System

A purely translational nodal displacement-based kinetic energy results in the equation of motion

$$\frac{d}{dt}\left(\frac{\partial T}{\partial\{\dot{\rho}\}}\right) - \frac{\partial T}{\partial\{\rho\}} = 0. \tag{1.52}$$

This is a subset of Lagrange's equation of motion, which will be expanded later. The derivative of the kinetic energy function obtained above with respect to the nodal displacement vector is

$$\frac{\partial T}{\partial\{\rho\}} = m(\Omega^2[J]\{r\} + \Omega^2[J]\{\rho\} + \Omega[P]\{\dot{\rho}\}). \tag{1.53}$$

The derivative with respect to the nodal velocity vector and the following time differentiation results in the equation

$$\frac{d}{dt}\left(\frac{\partial T}{\partial \{\dot{\rho}\}}\right) = m(\Omega[P]^T \{\dot{\rho}\} + [I]\{\ddot{\rho}\}).$$ (1.54)

Substituting these terms into Equation (1.52) and using the relation $-[P] = [P]^T$ yields

$$m(\Omega[P]^T \{\dot{\rho}\} + [I]\{\ddot{\rho}\} - \Omega^2[J]\{r\} - \Omega^2[J]\{\rho\} + \Omega[P]^T \{\dot{\rho}\}) = 0.$$ (1.55)

Collecting acceleration, velocity, and displacement terms, the equation of motion of the particle in terms of the nodal translational displacements can be written as

$$m[I]\{\ddot{\rho}\} + 2\Omega m[P]\{\dot{\rho}\} - \Omega^2 m[J]\{\rho\} = \Omega^2 m[J]\{r\}.$$ (1.56)

The right-hand side is the centrifugal force resulting from the rotation of the particle, which is dependent on the rotational speed and the radial location of the particle,

$$\Omega^2 m[J]\{r\} = m\Omega^2 \begin{bmatrix} 1 & 0 & 0 \\ 0 & 1 & 0 \\ 0 & 0 & 0 \end{bmatrix} \begin{Bmatrix} x \\ y \\ z \end{Bmatrix} = m\Omega^2 \begin{Bmatrix} x \\ y \\ 0 \end{Bmatrix} = \{f_{cp}\}.$$ (1.57)

We introduce new matrices. The first term is called the mass matrix:

$$[M_\rho] = m[I] = \begin{bmatrix} m & & \\ & m & \\ & & m \end{bmatrix}.$$ (1.58)

The second term results in the Coriolis term of the gyroscopic matrix:

$$[C_\rho] = m[P]^T = \begin{bmatrix} 0 & -m & 0 \\ m & 0 & 0 \\ 0 & 0 & 0 \end{bmatrix}.$$ (1.59)

Finally, the third term's coefficient produces the so-called centrifugal soften-ing matrix resulting from the nodal displacement of the particle of the form

$$[Z_\rho] = m[J] = \begin{bmatrix} m & & \\ & m & \\ & & 0 \end{bmatrix}. \tag{1.60}$$

The matrix equation of motion of a rotating particle in a rotating coordinate system, due to the kinetic energy only and in terms of the translational dis-placement, becomes

$$[M_\rho]\{\ddot{\rho}\} + 2\Omega[C_\rho]\{\dot{\rho}\} - \Omega^2[Z_\rho]\{\rho\} = \{f_{c\rho}\}. \tag{1.61}$$

The solution of this differential equation is the nodal translational displace-ment of the particle, presented in the rotational coordinate system.

A similar approach is used for the nodal rotational displacement-based kinetic energy derived in Equation (1.51). The derivative with respect to the velocity is

$$\frac{\partial T}{\partial\{\dot{\alpha}\}} = m(\Omega[A]^T[P]^T\{r\} + \Omega[A]^T[P]^T[A]\{\alpha\} + [A]^T[I][A]\{\dot{\alpha}\}), \tag{1.62}$$

and its temporal derivative yields

$$\frac{d}{dt}\frac{\partial T}{\partial\{\dot{\alpha}\}} = m(\Omega[A]^T[P]^T[A]\{\dot{\alpha}\} + [A]^T[I][A]\{\ddot{\alpha}\}). \tag{1.63}$$

The derivative with respect to the displacement is

$$\frac{\partial T}{\partial\{\alpha\}} = m(\Omega^2[A]^T[J]\{r\} + \Omega^2[A]^T[J][A]\{\alpha\} + \Omega[A]^T[P][A]\{\dot{\alpha}\}). \tag{1.64}$$

Subtracting them brings the Lagrange equation of motion for this case:

$$m([A]^T[I][A]\{\ddot{\alpha}\} + 2\Omega[A]^T[P]^T[A]\{\dot{\alpha}\} - \Omega^2[A]^T[J][A]\{\alpha\}$$
$$- \Omega^2[A]^T[J]\{r\}) = 0. \tag{1.65}$$

Note that in computing the second term, the relation $[P]^T = -[P]$ was used again and resulted in the coefficient 2. Let us now discuss each of these terms and derive their final forms. The first term produces the inertia force. The matrix product

$$
m[A]^T[A] = m \begin{bmatrix} 0 & -z' & y' \\ z' & 0 & -x' \\ -y' & x' & 0 \end{bmatrix} \begin{bmatrix} 0 & z' & -y' \\ -z' & 0 & x' \\ y' & -x' & 0 \end{bmatrix}
$$

$$
= m \begin{bmatrix} z'^2 + y'^2 & -x'y' & -x'z' \\ -x'y' & z'^2 + x'^2 & -y'z' \\ -x'z' & -y'z' & y'^2 + x'^2 \end{bmatrix} \tag{1.66}
$$

generates inertia terms. Considering the definition of the mass components described in the last section, the summations for the six submasses will result in the cancelation of the off-diagonal terms, producing

$$
\sum_{i=1}^{6} m_i [A]^T[A] = 2m' \begin{bmatrix} z'^2 + y'^2 & 0 & 0 \\ 0 & z'^2 + x'^2 & 0 \\ 0 & 0 & x'^2 + y'^2 \end{bmatrix}.
$$

Utilizing the moments of inertia of the mass particle with respect to the various coordinate axes as introduced in the last section,

$$
\Theta_x = 2m'(z'^2 + y'^2)
$$

$$
\Theta_y = 2m'(z'^2 + x'^2) \tag{1.67}
$$

$$
\Theta_z = 2m'(x'^2 + y'^2),
$$

the matrix of the inertia terms is the mass matrix

$$
[M_\alpha] = \begin{bmatrix} \Theta_x & 0 & 0 \\ 0 & \Theta_y & 0 \\ 0 & 0 & \Theta_z \end{bmatrix}. \tag{1.68}
$$

The second term in the Lagrange equation is the triple product

$$[A]^T [P]^T [A] = \begin{bmatrix} 0 & -z' & y' \\ z' & 0 & -x' \\ -y' & x' & 0 \end{bmatrix} \begin{bmatrix} 0 & -1 & 0 \\ 1 & 0 & 0 \\ 0 & 0 & 0 \end{bmatrix} \begin{bmatrix} 0 & z' & -y' \\ -z' & 0 & x' \\ y' & -x' & 0 \end{bmatrix}$$

$$= \begin{bmatrix} 0 & -z'^2 & y'z \\ z'^2 & 0 & -z'x' \\ -y'z' & x'z' & 0 \end{bmatrix}, \tag{1.69}$$

and the same cancelation occurs due to the signs of the mass components, hence

$$\sum_{i=1}^{6} m'[A]^T [P]^T [A] = 2m' \begin{bmatrix} 0 & -z'^2 & 0 \\ z'^2 & 0 & 0 \\ 0 & 0 & 0 \end{bmatrix}.$$

Using again the inertia relations of the last section

$$2m' \, z'^2 = 3m'\left(i_x^2 + i_y^2 + i_z^2\right) = \frac{1}{2}(\Theta_x + \Theta_y - \Theta_z), \tag{1.70}$$

the gyroscopic matrix will become:

$$[C_\alpha] = \begin{bmatrix} 0 & -\dfrac{1}{2}(\Theta_x + \Theta_y - \Theta_z) & 0 \\ \dfrac{1}{2}(\Theta_x + \Theta_y - \Theta_z) & 0 & 0 \\ 0 & 0 & 0 \end{bmatrix}. \tag{1.71}$$

The third term of the Lagrange equation is based on the triple product $[A]^T[J][A]$. Computing this matrix with using only the first order term in the cosine approximation would result in missing terms that are necessary for physics fidelity. The accurate form using the quadratic term is derived in the Appendix and is of the form:

$$\begin{bmatrix} y'^2 - z'^2 & 2x'y' & 2x'z' \\ 2x'y' & x'^2 - z'^2 & 2y'z' \\ 2x'z' & 2y'z' & \left(x'^2 + y'^2\right) - \left(x'^2 + y'^2\right) \end{bmatrix} \tag{1.72}$$

The derivation using the quadratic cosine approximation is very involved and since the other terms do not suffer from the first order cosine approximation it is not used in this section.

Summing the six sub-masses results in the cancellation of the cross products as

$$
2m' \begin{bmatrix} y'^2 - z'^2 & 0 & 0 \\ 0 & x'^2 - z'^2 & 0 \\ 0 & 0 & 0 \end{bmatrix}, \tag{1.73}
$$

and becomes the centrifugal softening matrix:

$$
[Z_\alpha] = \begin{bmatrix} \Theta_y - \Theta_z & 0 & 0 \\ 0 & \Theta_x - \Theta_z & 0 \\ 0 & 0 & 0 \end{bmatrix}. \tag{1.74}
$$

Finally, the fourth term is based on the product

$$
[A]^T[J] = \begin{bmatrix} 0 & -z' & y' \\ z' & 0 & -x' \\ -y' & x' & 0 \end{bmatrix} \begin{bmatrix} 1 & 0 & 0 \\ 0 & 1 & 0 \\ 0 & 0 & 0 \end{bmatrix} = \begin{bmatrix} 0 & -z' & 0 \\ z' & 0 & 0 \\ -y' & x' & 0 \end{bmatrix}, \tag{1.75}
$$

and the expression becomes the centrifugal force acting on the particle:

$$
m\Omega^2 \begin{bmatrix} 0 & -z' & 0 \\ z' & 0 & 0 \\ -y' & x' & 0 \end{bmatrix} \begin{Bmatrix} x' \\ y' \\ z' \end{Bmatrix} = m\Omega^2 \begin{Bmatrix} -z'y' \\ z'x' \\ 0 \end{Bmatrix} = \{f_{c\alpha}\}. \tag{1.76}
$$

The resulting equation of motion is structurally similar to Equation 1.61,

$$
[M_\alpha]\{\ddot{\alpha}\} + 2\Omega[C_\alpha]\{\dot{\alpha}\} - \Omega^2[Z_\alpha]\{\alpha\} = \{f_{c\alpha}\}, \tag{1.77}
$$

however, the content of the matrices is different. The solution of this differential equation produces the nodal rotational displacement of the rotating particle, also in the rotational coordinate system.

The equilibrium equation of a particle simultaneously undergoing translational and rotational displacement is obtained by simply assembling Equations (1.61) and (1.77) as

$$
\begin{bmatrix} [M_\rho] & 0 \\ 0 & [M_\alpha] \end{bmatrix} \begin{Bmatrix} \{\ddot{\rho}\} \\ \{\ddot{\alpha}\} \end{Bmatrix} + 2\Omega \begin{bmatrix} [C_\rho] & 0 \\ 0 & [C_\alpha] \end{bmatrix} \begin{Bmatrix} \{\dot{\rho}\} \\ \{\dot{\alpha}\} \end{Bmatrix}
$$

$$
-\Omega^2 \begin{bmatrix} [Z_\rho] & 0 \\ 0 & [Z_\alpha] \end{bmatrix} \begin{Bmatrix} \{\rho\} \\ \{\alpha\} \end{Bmatrix} = \begin{Bmatrix} \{f_{c\rho}\} \\ \{f_{c\alpha}\} \end{Bmatrix}. \tag{1.78}
$$

This equation does not yet consider the coupling of the local translations and rotations; that subject is addressed in the next chapter. It is important to notice, however, that in the rotating system both the nodal translations and the rotations contribute to the rotational matrices, making it a generic analysis scenario. There are also some merits in executing analysis, or at least interpreting analysis results, in a fixed reference system; hence the next section will focus on that.

1.8 Equation of Motion in the Fixed Coordinate System

The above equilibrium equations were written in terms of the particle's nodal displacement in the rotating coordinate system. In some cases it is advantageous to execute the analysis in the fixed system, that is, define the particle's generalized coordinates in the fixed coordinate system. The procedure is similar to the above; however, the location vector changes. For clarity and distinction, we will recall the earlier case before we introduce the new one. The location vector in the case of a rotating system for translational nodal displacement was

$$
\{\bar{r}\} = [H](\{r\} + \{\rho\}), \tag{1.79}
$$

and now in the case of the fixed system it becomes

$$
\{\bar{r}\} = [H]\{r\} + \{\bar{\rho}\}. \tag{1.80}
$$

Here the $\{\bar{\rho}\}$ represents the displacement of the point with respect to the fixed coordinate system due to the nodal translational displacement and not subject to the rotational matrix multiplication. Hence the velocity formula is simpler in the fixed system:

$$\{\dot{\bar{r}}\} = \Omega[\bar{H}]\{r\} + \{\dot{\rho}\}. \tag{1.81}$$

This simplicity carries into the kinetic energy that in this case becomes

$$T = \frac{m}{2}(\Omega\{r\}^T[\bar{H}]^T + \{\dot{\rho}\}^T)(\Omega[\bar{H}]\{r\} + \{\dot{\rho}\}). \tag{1.82}$$

Executing the posted multiplications as

$$T = \frac{m}{2}(\Omega^2\{r\}^T[\bar{H}]^T[\bar{H}]\{r\} + \Omega\{r\}^T[\bar{H}]^T\{\dot{\rho}\} + \Omega\{\dot{\rho}\}^T[\bar{H}]\{r\} + \{\dot{\rho}\}^T\{\dot{\rho}\}) \tag{1.83}$$

and collecting the terms results in

$$T = \frac{m}{2}(\Omega^2\{r\}^T[\bar{H}]^T[\bar{H}]\{r\} + 2\Omega\{\dot{\rho}\}^T[\bar{H}]\{r\} + \{\dot{\rho}\}^T\{\dot{\rho}\}). \tag{1.84}$$

Finally, using the earlier introduced matrix products, the kinetic energy is

$$T = \frac{m}{2}(\Omega^2\{r\}^T[J]\{r\} + 2\Omega\{\dot{\rho}\}^T[\bar{H}]\{r\} + \{\dot{\rho}\}^T\{\dot{\rho}\}). \tag{1.85}$$

The simplified Lagrange's equation of motion for this case is

$$\frac{d}{dt}\left(\frac{\partial T}{\partial\{\dot{\rho}\}}\right) - \frac{\partial T}{\partial\{\rho\}} = 0. \tag{1.86}$$

Because the first term of the kinetic energy is constant and there is no term dependent on $\{\rho\}$, the equation of motion for this case becomes the rather simple

$$m[I]\{\ddot{\rho}\} - m\Omega[\dot{\bar{H}}]\{r\} = 0. \tag{1.87}$$

Computing the derivative as

$$[\dot{\bar{H}}] = \Omega\begin{bmatrix} -\cos\Omega t & \sin\Omega t & 0 \\ -\sin\Omega t & -\cos\Omega t & 0 \\ 0 & 0 & 0 \end{bmatrix}, \tag{1.88}$$

the second term becomes the centrifugal force:

$$m\Omega[\dot{H}]\{r\} = m\Omega^2 \begin{bmatrix} -\cos\Omega t & \sin\Omega t & 0 \\ -\sin\Omega t & -\cos\Omega t & 0 \\ 0 & 0 & 0 \end{bmatrix} \begin{Bmatrix} x \\ y \\ z \end{Bmatrix}$$

$$= m\Omega^2 \begin{Bmatrix} -x\cos\Omega t + y\sin\Omega t \\ -x\sin\Omega t - y\cos\Omega t \\ 0 \end{Bmatrix} = \{\bar{f}_{cp}\}. \tag{1.89}$$

The bar indicates that it is a force with respect to the fixed system, and it is important to notice that the centrifugal force is periodic, that is, rotating in the fixed system. Introducing again

$$[M_\rho] = m[I] = \begin{bmatrix} m & & \\ & m & \\ & & m \end{bmatrix}, \tag{1.90}$$

the equation of motion of the particle due to the translational displacement-based kinetic energy is simply

$$[M_\rho]\{\ddot{\bar{\rho}}\} = \{\bar{f}_{cp}\}. \tag{1.91}$$

Solving the second-order differential equation will result in the displacement of the rotating particle in the fixed coordinate system.

In order to compute the fixed system equilibrium due to nodal rotation-based kinetic energy, we consider first that the particle undergoes the global rotation as

$$[H]\{r\}.$$

This will be followed by a local rotation with respect to the fixed system, and the location vector of the point with respect to the fixed system becomes

$$\{\bar{r}\} = [\tilde{A}][H]\{r\}, \tag{1.92}$$

where a transformation matrix with respect to the three fixed system coordinate axes is

$$[\tilde{A}] = \begin{bmatrix} 0 & -\bar{\theta} & \bar{\psi} \\ \bar{\theta} & 0 & -\bar{\varphi} \\ -\bar{\psi} & \bar{\varphi} & 0 \end{bmatrix}. \tag{1.93}$$

The nodal rotations with respect to the fixed system may be given in the vector

$$\{\bar{\alpha}\} = \begin{Bmatrix} \bar{\varphi} \\ \bar{\psi} \\ \bar{\theta} \end{Bmatrix}. \tag{1.94}$$

Let us first evaluate the right multiplication as

$$[H]\{r\} = \begin{bmatrix} x\cos\Omega t - y\sin\Omega t \\ x\sin\Omega t + y\cos\Omega t \\ z \end{bmatrix} = \begin{bmatrix} \bar{x}(\Omega t) \\ \bar{y}(\Omega t) \\ \bar{z} \end{bmatrix}.$$

Then the left multiplication is posted as

$$[\tilde{A}][H]\{r\} = \begin{bmatrix} 0 & -\bar{\theta} & \bar{\psi} \\ \bar{\theta} & 0 & -\bar{\varphi} \\ -\bar{\psi} & \bar{\varphi} & 0 \end{bmatrix} \begin{bmatrix} \bar{x}(\Omega t) \\ \bar{y}(\Omega t) \\ \bar{z} \end{bmatrix}.$$

Following the logic in the earlier section, we can write

$$[\tilde{A}][H]\{r\} = [\bar{A}]\{\bar{\alpha}\}. \tag{1.95}$$

Here the $[\bar{A}]$ matrix, while obtained with the same procedure as the $[A]$ matrix before, is written in terms of the mass point coordinates with respect to the fixed system, and it is also time dependent:

$$[\bar{A}](\Omega t) = \begin{bmatrix} 0 & \bar{z} & -\bar{y}(\Omega t) \\ -\bar{z} & 0 & \bar{x}(\Omega t) \\ \bar{y}(\Omega t) & -\bar{x}(\Omega t) & 0 \end{bmatrix}. \tag{1.96}$$

With this the location vector becomes

$$\{\bar{r}\} = [\bar{A}](\Omega t)\{\bar{\alpha}\}. \tag{1.97}$$

Mindful of the time dependence of the A matrix, the velocity in this scenario is

$$\{\dot{\bar{r}}\} = [\dot{\bar{A}}]\{\bar{\alpha}\} + [\bar{A}]\{\dot{\bar{\alpha}}\}. \tag{1.98}$$

The kinetic energy computation follows the earlier process of

$$T = \frac{m}{2}(\{\bar{\alpha}\}^T[\dot{\bar{A}}]^T + \{\dot{\alpha}\}^T[\bar{A}]^T)([\dot{\bar{A}}]\{\bar{\alpha}\} + [\bar{A}]\{\dot{\alpha}\}).$$ (1.99)

Executing the posted multiplications and collecting like terms results in

$$T = \frac{m}{2}(\{\bar{\alpha}\}^T[\dot{\bar{A}}]^T[\dot{\bar{A}}]\{\bar{\alpha}\} + 2\{\bar{\alpha}\}^T[\dot{\bar{A}}]^T[\bar{A}]\{\dot{\alpha}\} + \{\dot{\alpha}\}^T[\bar{A}]^T[\bar{A}]\{\dot{\alpha}\}).$$ (1.100)

The derivative with respect to the rotational displacement, after cancelation by 2, is

$$\frac{\partial T}{\partial\{\bar{\alpha}\}} = m([\dot{\bar{A}}]^T[\dot{\bar{A}}]\{\bar{\alpha}\} + [\dot{\bar{A}}]^T[\bar{A}]\{\dot{\alpha}\}).$$ (1.101)

The derivative with respect to the velocity, again by canceling the 2-s, is

$$\frac{\partial T}{\partial\{\dot{\bar{\alpha}}\}} = m([\bar{A}]^T[\dot{\bar{A}}]\{\bar{\alpha}\} + [\bar{A}]^T[\bar{A}]\{\dot{\alpha}\}).$$ (1.102)

Executing the time derivative we obtain

$$\frac{d}{dt}\frac{\partial T}{\partial\{\dot{\bar{\alpha}}\}} = m([\dot{\bar{A}}]^T[\dot{\bar{A}}]\{\bar{\alpha}\} + [\bar{A}]^T[\dot{\bar{A}}]\{\dot{\alpha}\} + [\bar{A}]^T[\dot{\bar{A}}]\{\bar{\alpha}\}$$

$$+ [\bar{A}]^T[\dot{\bar{A}}]\{\dot{\alpha}\} + [\bar{A}]^T[\bar{A}]\{\ddot{\alpha}\}).$$ (1.103)

Note that the possible second-order derivatives of the $[\bar{A}]$ matrix are not computed. The simplified (purely kinetic energy-based) Lagrange's equation in this case is

$$\frac{d}{dt}\left(\frac{\partial T}{\partial\{\dot{\alpha}\}}\right) - \frac{\partial T}{\partial\{\alpha\}} = 0.$$ (1.104)

The first and third terms of Equation (1.103) cancel out with the terms of Equation (1.101). The equation of motion with the remaining terms is

$$m(2[\bar{A}]^T[\dot{\bar{A}}]\{\dot{\alpha}\} + [\bar{A}]^T[\bar{A}]\{\ddot{\alpha}\}) = 0.$$ (1.105)

The derivative in the first term is computed as

$$[\dot{\bar{A}}] = \Omega \begin{bmatrix} 0 & 0 & -\dot{\bar{y}} \\ 0 & 0 & \dot{\bar{x}} \\ \dot{\bar{y}} & -\dot{\bar{x}} & 0 \end{bmatrix} = \Omega \begin{bmatrix} 0 & 0 & -\bar{x} \\ 0 & 0 & -\bar{y} \\ \bar{x} & \bar{y} & 0 \end{bmatrix}. \tag{1.106}$$

Evaluating the first term results in $2[\bar{A}][\dot{\bar{A}}] = \Omega[C]_{\bar{\alpha}}$, where the gyroscopic matrix

$$\begin{bmatrix} 0 & -\bar{\Theta}_z & 0 \\ \bar{\Theta}_z & 0 & 0 \\ 0 & 0 & 0 \end{bmatrix} = [C_{\bar{\alpha}}] \tag{1.107}$$

was computed by omitting the cross products of inertia ($\overline{xy}$ type) terms results and assuming symmetric rotors, $\Theta_x = \Theta_y$. Note that the moments of inertia are with respect to the fixed system coordinate axes, as denoted by the bar.

The second term, based on

$$[\bar{A}]^T[\bar{A}] = \begin{bmatrix} \bar{y}^2 + \bar{z}^2 & 0 & 0 \\ 0 & \bar{x}^2 + \bar{z}^2 & 0 \\ 0 & 0 & \bar{y}^2 + \bar{x}^2 \end{bmatrix}, \tag{1.108}$$

produces the mass matrix of inertia terms as

$$m[\bar{A}]^T[\bar{A}] = \begin{bmatrix} \bar{\Theta}_x & & 0 \\ & \bar{\Theta}_y & 0 \\ 0 & 0 & \bar{\Theta}_z \end{bmatrix} = [M_{\bar{\alpha}}]. \tag{1.109}$$

The cross products of inertia are omitted, as before. The equilibrium equation of motion in the fixed system, due to rotational nodal displacements, becomes

$$[M_{\bar{\alpha}}]\{\ddot{\bar{\alpha}}\} + \Omega[C_{\bar{\alpha}}]\{\dot{\bar{\alpha}}\} = 0. \tag{1.110}$$

Again, by simply assembling Equations (1.91) and (1.110), we can see the simultaneous local translations and rotations in the fixed system as

$$
\begin{bmatrix} [M_{\bar{\rho}}] & 0 \\ 0 & [M_{\bar{\alpha}}] \end{bmatrix} \begin{Bmatrix} \{\ddot{\bar{\rho}}\} \\ \{\ddot{\bar{\alpha}}\} \end{Bmatrix} + \Omega \begin{bmatrix} 0 & 0 \\ 0 & [C_{\bar{\alpha}}] \end{bmatrix} \begin{Bmatrix} \{\ddot{\bar{\rho}}\} \\ \{\dot{\bar{\alpha}}\} \end{Bmatrix} = \begin{Bmatrix} \{f_{c\bar{\rho}}\} \\ \{0\} \end{Bmatrix}.
$$

(1.111)

This equation still does not consider the coupling of the local translations and rotations. It is also much simpler than the concluding equation of the last section, but that is deceiving. In the fixed system equilibrium, only the nodal rotations contributed to the gyroscopic matrix, and the nodal displacements simply described Newton's second law. This simplicity actually limits the types of analyses one can execute in the fixed system, a topic discussed in more detail in Chapter 6, Section 6.1.

2

Coupled Solution Formulations

In Chapter 1, we developed a methodology to represent two distinct nodal displacements of the particles of a rotating structure. In this chapter we focus on coupling the rotating part's displacements to each other as well as to the stationary part of the structure.

First, we will present a matrix formulation of Lagrange's equations of motion to aid the development of various coupled cases. Then we will address the topic of coupling the nodal translations of the rotating model to the stationary world. The simultaneous coupling of both types of nodal displacements follows.

Finally, the coupling to the stationary part of the structure is discussed. After all, apart from some space satellites rotating by themselves in space, without a connection to a fixed environment, all Earth-bound rotating equipments are coupled with the fixed environment.

The intermediate scenarios are presented because in some applications they are adequate to capture the physics. The chapter concludes with a discussion of the time-dependent terms arising in the coupled scenarios.

2.1 Matrix Formulation of Lagrange's Equations

The various coupling scenarios will be discussed based on a common matrix formulation of the location vector with respect to the stationary (fixed) system and its time derivative as follows:

$$\{\bar{r}\} = [M]\{g\}; \quad \{\dot{\bar{r}}\} = [M]\{\dot{g}\} + [\dot{M}]\{g\}.$$

The generalized coordinate vector $\{g\}$ and the governing matrix $[M]$ will vary in the different scenarios. The kinetic energy is

$$T = \frac{1}{2}m([M]\{\dot{g}\} + [\dot{M}]\{g\})^T([M]\{\dot{g}\} + [\dot{M}]\{g\})$$

$$= \frac{1}{2}m(\{\dot{g}\}^T[M]^T[M]\{\dot{g}\} + \{g\}^T[\dot{M}]^T[\dot{M}]\{g\} + 2\{g\}^T[\dot{M}]^T[M]\{\dot{g}\}). \quad (2.1)$$

The derivatives required for Lagrange's equation of motion become

$$\frac{\partial}{\partial\{g\}}T = m([\dot{M}]^T[\dot{M}]\{g\} + [\dot{M}]^T[M]\{\dot{g}\}),$$

$$\frac{\partial}{\partial\{\dot{g}\}}T = m([M]^T[M]\{\dot{g}\} + [M]^T[\dot{M}]\{g\}),$$

and

$$\frac{d}{dt}\frac{\partial}{\partial\{\dot{g}\}}T = m([M]^T[M]\{\ddot{g}\} + 3[M]^T[\dot{M}]\{\dot{g}\} + [\dot{M}]^T[\dot{M}]\{g\} + [M]^T[\ddot{M}]\{g\}).$$

After the cancellations, Lagrange's equation of motion is of the form

$$\frac{d}{dt}\frac{\partial}{\partial\{\dot{g}\}}T - \frac{\partial}{\partial\{g\}}T = m([M]^T[M]\{\ddot{g}\} + 2[M]^T[\dot{M}]\{\dot{g}\} + [M]^T[\ddot{M}]\{g\}) = 0. \quad (2.2)$$

This formulation, in connection with the specific matrix of a certain scenario and the corresponding generalized coordinate vector, is used in the following sections.

2.2 Coupling Nodal Translations to the Stationary Part

In Chapter 1, we restricted the offset between the origins of the fixed and the rotating systems to be zero. We now relax the restriction on the vector $\{\sigma\}$ between the coordinate systems in order to accomplish the goal of this section, coupling the stationary and rotating parts. The components of this vector are

$$\{\sigma\} = \begin{Bmatrix} \bar{u} \\ \bar{v} \\ \bar{w} \end{Bmatrix}. \quad (2.3)$$

We present the scenario in Figure 2.1.

We are going to focus on the nodal translational displacement $\{\rho\}$ of the mass particle in this section. We will allow nodal rotations also in the next section and later derive the generic case as well. The location vector in this scenario is

$$\{\bar{r}\} = \{\sigma\} + [H](\{r\} + \{\rho\}) = \{\sigma\} + [H]\{\rho\} + [H]\{r\}. \quad (2.4)$$

We turn to the matrix formulation introduced in the last section:

$$\{\bar{r}\} = [M]\{g\}.$$

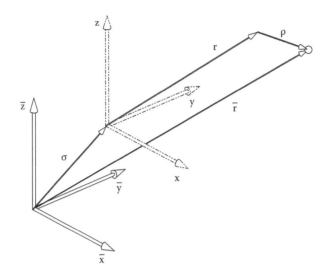

FIGURE 2.1
Coupling nodal translations to the stationary part.

The matrix governing this scenario is

$$[M] = \begin{bmatrix} [I] & [H] & [H] \end{bmatrix},$$

(2.5)

and the augmented generalized coordinate vector is

$$\{g\} = \begin{Bmatrix} \{\sigma\} \\ \{\rho\} \\ \{r\} \end{Bmatrix}.$$

(2.6)

The three matrices participating in the Lagrange equation for this scenario are given by the dyadic products:

$$[M]^T[M] = \begin{bmatrix} [I] \\ [H]^T \\ [H]^T \end{bmatrix} \begin{bmatrix} [I] & [H] & [H] \end{bmatrix} = \begin{bmatrix} [I] & [H] & [H] \\ [H]^T & [I] & [I] \\ [H]^T & [I] & [I] \end{bmatrix},$$

$$[M]^T[\dot{M}] = \Omega \begin{bmatrix} [I] \\ [H]^T \\ [H]^T \end{bmatrix} \begin{bmatrix} [0] & [\bar{H}] & [\bar{H}] \end{bmatrix} = \Omega \begin{bmatrix} [0] & [\bar{H}] & [\bar{H}] \\ [0] & [P]^T & [P]^T \\ [0] & [P]^T & [P]^T \end{bmatrix},$$

and

$$[M]^T[\ddot{M}] = \Omega^2 \begin{bmatrix} [I] \\ [H]^T \\ [H]^T \end{bmatrix} \begin{bmatrix} [0] & [\bar{H}] & [\bar{\bar{H}}] \end{bmatrix} = \Omega^2 \begin{bmatrix} [0] & [\bar{H}] & [\bar{\bar{H}}] \\ [0] & -[J] & -[J] \\ [0] & -[J] & -[J] \end{bmatrix}.$$

The matrix form of Lagrange's equation for this scenario becomes

$$m \begin{bmatrix} [I] & [H] & [H] \\ [H]^T & [I] & [I] \\ [H]^T & [I] & [I] \end{bmatrix} \begin{Bmatrix} \{\ddot{\sigma}\} \\ \{\ddot{\rho}\} \\ \{0\} \end{Bmatrix} + 2m\Omega \begin{bmatrix} [0] & [\bar{H}] & [\bar{\bar{H}}] \\ [0] & [P]^T & [P]^T \\ [0] & [P]^T & [P]^T \end{bmatrix} \begin{Bmatrix} \{\dot{\sigma}\} \\ \{\dot{\rho}\} \\ \{0\} \end{Bmatrix} +$$

$$+ m\Omega^2 \begin{bmatrix} [0] & [\bar{H}] & [\bar{\bar{H}}] \\ [0] & -[J] & -[J] \\ [0] & -[J] & -[J] \end{bmatrix} \begin{Bmatrix} \{\sigma\} \\ \{\rho\} \\ \{r\} \end{Bmatrix} = 0. \tag{2.7}$$

Because the derivatives of the $\{r\}$ vector vanish, the first column and the last row may be partitioned and the nonvanishing terms moved to the right-hand side, resulting in the equilibrium equation

$$m \begin{bmatrix} [I] & [H] \\ [H]^T & [I] \end{bmatrix} \begin{Bmatrix} \{\ddot{\sigma}\} \\ \{\ddot{\rho}\} \end{Bmatrix} + 2m\Omega \begin{bmatrix} 0 & [\bar{H}] \\ 0 & [P]^T \end{bmatrix} \begin{Bmatrix} \{\dot{\sigma}\} \\ \{\dot{\rho}\} \end{Bmatrix} + m\Omega^2 \begin{bmatrix} 0 & [\bar{H}] \\ 0 & -[J] \end{bmatrix} \begin{Bmatrix} \{\sigma\} \\ \{\rho\} \end{Bmatrix}$$

$$= m\Omega^2 \begin{bmatrix} -[\bar{\bar{H}}] \\ [J] \end{bmatrix} \{r\}. \tag{2.8}$$

Using earlier established relations for the new partition, this specific case of the coupled equilibrium becomes

$$\begin{bmatrix} [M_\sigma] & m[H] \\ m[H]^T & [M_\rho] \end{bmatrix} \begin{Bmatrix} \{\ddot{\sigma}\} \\ \{\ddot{\rho}\} \end{Bmatrix} + 2\Omega \begin{bmatrix} 0 & m[\bar{H}] \\ 0 & [C_\rho] \end{bmatrix} \begin{Bmatrix} \{\dot{\sigma}\} \\ \{\dot{\rho}\} \end{Bmatrix} + \Omega^2 \begin{bmatrix} 0 & m[\bar{H}] \\ 0 & -[Z_\rho] \end{bmatrix} \begin{Bmatrix} \{\sigma\} \\ \{\rho\} \end{Bmatrix}$$

$$= \begin{bmatrix} -m\Omega^2[\bar{\bar{H}}]\{r\} \\ \{f_{cp}\} \end{bmatrix}. \tag{2.9}$$

Note that the matrices $[H]$, $[\bar{H}]$, and $[\bar{\bar{H}}]$ are time dependent; hence this equation is periodic in nature. We will discuss the computation of such terms in Section 2.5 and the solution of periodic systems in Chapter 5. Section 5.4.

2.3 Simultaneous Coupling of Translations and Rotations

The next coupling scenario we consider is the simultaneous coupling of nodal rotations and translations with the stationary part shown in Figure 2.2.

The location vector of this scenario is of the form

$$\{\bar{r}\} = \{\sigma\} + [H](\{r\} + \{\rho\} + [A]\{\alpha\}) = \{\sigma\} + [H]\{r\} + [H]\{\rho\} + [H][A]\{\alpha\}. \quad (2.10)$$

The first term is the offset from the stationary system to the rotating coordinate system. The second term is the location vector of the particle given in the rotating system, but transformed to the fixed system. The third term is the nodal translation vector of the particle, and the last term is the translational equivalent of the nodal rotation. Both of these are also transformed to the fixed system. The now familiar matrix formulation is

$$\{\bar{r}\} = [M]\{g\}.$$

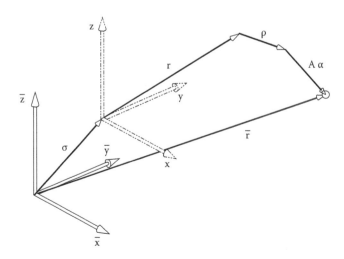

FIGURE 2.2
Simultaneous coupling to the stationary part.

This scenario has the governing matrix

$$[M] = \begin{bmatrix} [I] & [H] & [H][A] & [H] \end{bmatrix}$$ (2.11)

and the generalized coordinate vector of

$$\{g\} = \begin{Bmatrix} \{\sigma\} \\ \{\rho\} \\ \{\alpha\} \\ \{r\} \end{Bmatrix}.$$ (2.12)

The three matrices participating in the Lagrange equation for this scenario are

$$[M]^T[M] = \begin{bmatrix} [I] \\ [H]^T \\ [A]^T[H]^T \\ [H]^T \end{bmatrix} \begin{bmatrix} [I] & [H] & [H][A] & [H] \end{bmatrix}$$

$$= \begin{bmatrix} [I] & [H] & [H][A] & [H] \\ [H]^T & [I] & [A] & [I] \\ [A]^T[H]^T & [A]^T & [A]^T[A] & [A]^T \\ [H]^T & [I] & [A] & [I] \end{bmatrix},$$

$$[M]^T[\dot{M}] = \Omega \begin{bmatrix} [I] \\ [H]^T \\ [A]^T[H]^T \\ [H]^T \end{bmatrix} \begin{bmatrix} [0] & [\bar{H}] & [\bar{H}][A] & [\bar{H}] \end{bmatrix}$$

$$= \Omega \begin{bmatrix} [0] & [\bar{H}] & [\bar{H}][A] & [\bar{H}] \\ [0] & [P]^T & [P]^T[A] & [P]^T \\ [0] & [A]^T[P]^T & [A]^T[P]^T[A] & [A]^T[P]^T \\ [0] & [P]^T & [P]^T[A] & [P]^T \end{bmatrix},$$

and

$$[M]^T[\ddot{M}] = \Omega^2 \begin{bmatrix} [I] \\ [H]^T \\ [A]^T[H]^T \\ [H]^T \end{bmatrix} \begin{bmatrix} [0] & [\bar{\bar{H}}] & [\bar{\bar{H}}][A] & [\bar{\bar{H}}] \end{bmatrix}$$

$$= \Omega^2 \begin{bmatrix} [0] & [\bar{\bar{H}}] & [\bar{\bar{H}}][A] & [\bar{\bar{H}}] \\ [0] & -[J] & -[J][A] & -[J] \\ [0] & -[A]^T[J] & -[A]^T[J][A] & -[A]^T[J] \\ [0] & -[J] & -[J][A] & -[J] \end{bmatrix}.$$

The relations $[H]^T[\bar{H}] = [P]^T$ and $[H]^T[\bar{\bar{H}}] = -[J]$ are used in the above matrices. The combined system for this scenario becomes

$$m \begin{bmatrix} [I] & [H] & [H][A] \\ [H]^T & [I] & [A] \\ [A]^T[H]^T & [A]^T & [A]^T[A] \end{bmatrix} \begin{Bmatrix} \{\ddot{\sigma}\} \\ \{\ddot{\rho}\} \\ \{\ddot{\alpha}\} \end{Bmatrix} + 2m\Omega \begin{bmatrix} [0] & [\bar{H}] & [\bar{H}][A] \\ [0] & [P]^T & [P]^T[A] \\ [0] & [A]^T[P]^T & [A]^T[P]^T[A] \end{bmatrix}$$

$$\times \begin{Bmatrix} \{\dot{\sigma}\} \\ \{\dot{\rho}\} \\ \{\dot{\alpha}\} \end{Bmatrix} + m\Omega^2 \begin{bmatrix} [0] & [\bar{\bar{H}}] & [\bar{\bar{H}}][A] \\ [0] & -[J] & -[J][A] \\ [0] & -[A]^T[J] & -[A]^T[J][A] \end{bmatrix} \begin{Bmatrix} \{\sigma\} \\ \{\rho\} \\ \{\alpha\} \end{Bmatrix} = m\Omega^2 \begin{bmatrix} -[\bar{\bar{H}}] \\ [J] \\ [A]^T[J] \end{bmatrix} \{r\}.$$

$$(2.13)$$

If one uses the simplified nodal mass and inertia formulation introduced in Chapter 1, Section 1.6, the terms $m[A], m[P][A], m[J][A]$ will vanish. If the nodal inertia information is provided directly by the user of software, or by a more elaborate scheme reflecting the mesh environment surrounding the node point, they may not vanish. Hence in the following we'll retain these terms for the sake of generality.

By reusing the earlier introduced matrices of

$$m[I] = M_\rho, \ m[A]^T[A] = M_\alpha, \ m[P]^T = [C_\rho], \ m[A]^T[P]^T[A] = [C_\alpha]$$

and

$$m[J] = [Z_\rho], \ m[A]^T[J][A] = [Z_\alpha], \ m\Omega^2[A]^T[J]\{r\} = [f_{c\alpha}], \ m\Omega^2[J]\{r\} = [f_{cp}],$$

the coupled equation is of the final form

$$
\begin{bmatrix} [M_\sigma] & m[H] & m[H][A] \\ m[H]^T & [M_\rho] & m[A] \\ m[A]^T[H]^T & m[A]^T & [M_\alpha] \end{bmatrix} \begin{Bmatrix} \{\ddot{\sigma}\} \\ \{\ddot{\rho}\} \\ \{\ddot{\alpha}\} \end{Bmatrix} + 2\Omega \begin{bmatrix} [0] & m[\bar{H}] & m[\bar{H}][A] \\ [0] & [C_\rho] & m[P]^T[A] \\ [0] & m[A]^T[P]^T & [C_\alpha] \end{bmatrix}
$$

$$
\times \begin{Bmatrix} \{\dot{\sigma}\} \\ \{\dot{\rho}\} \\ \{\dot{\alpha}\} \end{Bmatrix} + \Omega^2 \begin{bmatrix} [0] & m[\bar{\bar{H}}] & m[\bar{\bar{H}}][A] \\ [0] & [Z_\rho] & -m[J][A] \\ [0] & -m[A]^T[J] & [Z_\alpha] \end{bmatrix} \begin{Bmatrix} \{\sigma\} \\ \{\rho\} \\ \{\alpha\} \end{Bmatrix} = \begin{bmatrix} -m\Omega^2[\bar{\bar{H}}]\{r\} \\ \{f_{cp}\} \\ \{f_{c\alpha}\} \end{bmatrix}. \quad (2.14)
$$

Because the matrices $[H], [\bar{H}], [\bar{\bar{H}}]$ are time dependent, this equation is also periodic.

2.4 Full Coupling of the Stationary and Rotating Parts

Thus far we have only considered a translation between the stationary and the rotating parts, representing only a displacement of the rotor with respect to the stationary part. Allowing a small rotation of the rotating frame of reference is also necessary in industrial practice. This would represent the tilting of the stationary part, which occurs, for example, when the tower supporting the wind turbine undergoes bending—a topic given more attention in Chapter 10.

This scenario is depicted in Figure 2.3, where the small rotation of the rotating frame of reference is depicted by the nonparallel coordinate axes.

Let us define the small rotations of the rotor reference point (the origin of the rotating system) along the three coordinate axes as

$$
\{\beta\} = \begin{Bmatrix} \bar{\varphi} \\ \bar{\psi} \\ \bar{\theta} \end{Bmatrix}. \quad (2.15)
$$

The notation reflects the fact that the angles are with respect to the stationary coordinate system, but otherwise the same convention applies. Following the procedure developed in Chapter 1, Section 1.6 regarding the small nodal rotations, the transformation matrix becomes

$$
[B] = \begin{bmatrix} 0 & -\bar{\theta} & \bar{\psi} \\ \bar{\theta} & 0 & -\bar{\varphi} \\ -\bar{\psi} & \bar{\varphi} & 0 \end{bmatrix}. \quad (2.16)
$$

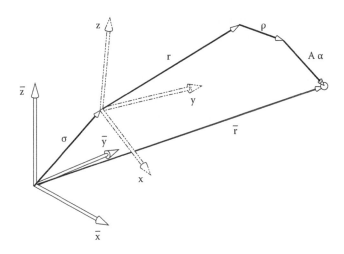

FIGURE 2.3
Fully coupled stationary and rotating parts.

The relationship between the fixed system and the rotating system location vector of a point due to the reference point rotation becomes

$$\{\tilde{r}\} = \begin{bmatrix} 1 & -\bar{\theta} & \bar{\psi} \\ \bar{\theta} & 1 & -\bar{\phi} \\ -\bar{\psi} & \bar{\phi} & 1 \end{bmatrix} \{r\} = ([I] + [B])\{r\}. \tag{2.17}$$

The location vector of this most generic scenario is of the form

$$\{\bar{r}\} = \{\sigma\} + ([I] + [B])[H](\{r\} + \{\rho\} + [A]\{\alpha\}). \tag{2.18}$$

The identity matrix represents the case of Section 2.3, and the newly introduced term is due to the rotation of the reference point manifesting the behavior of a tilting stationary part. In order to follow the steps of the prior sections, we reorder as

$$\{\bar{r}\} = \{\sigma\} + [B][H]\{r\} + ([I] + [B])[H]\{\rho\} + ([I] + [B])[H][A]\{\alpha\}) + [H]\{r\}. \tag{2.19}$$

The terms involving the product of the B matrix and the displacements are quadratic in nature because the B matrix implicitly also contains displacements. These will be ignored in the proceeding, and the form of

$$\{\bar{r}\} = \{\sigma\} + [B][H]\{r\} + [H]\{\rho\} + [H][A]\{\alpha\}) + [H]\{r\} \tag{2.20}$$

will be basis of further developments.

This form does not explicitly contain the newly introduced rotations. The second term in the equation may be reorganized as

$$
[B][H]\{r\} = \begin{bmatrix} 0 & -\bar{\theta} & \bar{\psi} \\ \bar{\theta} & 0 & -\bar{\phi} \\ -\bar{\psi} & \bar{\phi} & 0 \end{bmatrix} \begin{bmatrix} \cos\Omega t & -\sin\Omega t & 0 \\ \sin\Omega t & \cos\Omega t & 0 \\ 0 & 0 & 1 \end{bmatrix} \begin{Bmatrix} x \\ y \\ z \end{Bmatrix}
$$

$$
= \begin{bmatrix} 0 & -\bar{\theta} & \bar{\psi} \\ \bar{\theta} & 0 & -\bar{\phi} \\ -\bar{\psi} & \bar{\phi} & 0 \end{bmatrix} \begin{Bmatrix} x\cos\Omega t - y\sin\Omega t \\ x\sin\Omega t + y\cos\Omega t \\ z \end{Bmatrix}
$$

$$
= \begin{Bmatrix} -\bar{\theta}(x\sin\Omega t + y\cos\Omega t) \\ \bar{\theta}(x\cos\Omega t - y\sin\Omega t) \\ -\bar{\psi}(x\cos\Omega t - y\sin\Omega t) + \bar{\phi}(x\sin\Omega t + y\cos\Omega t) \end{Bmatrix}
$$

$$
= \begin{bmatrix} 0 & 0 & -x\sin\Omega t - y\cos\Omega t \\ 0 & 0 & x\cos\Omega t - y\sin\Omega t \\ x\sin\Omega t + y\cos\Omega t & -x\cos\Omega t + y\sin\Omega t & 0 \end{bmatrix} \begin{Bmatrix} \bar{\phi} \\ \bar{\psi} \\ \bar{\theta} \end{Bmatrix}
$$

$$
= [B_0]\{\beta\}.
$$

$$(2.21)$$

The augmented generalized coordinate vector is of the form

$$
\{g\} = \begin{Bmatrix} \{\sigma\} \\ \{\beta\} \\ \{\rho\} \\ \{\alpha\} \\ \{r\} \end{Bmatrix},
$$

and the governing matrix becomes

$$
[M] = \begin{bmatrix} [I] & [B_0] & [H] & [H][A] & [H] \end{bmatrix}.
$$

$$(2.22)$$

With $[\dot{H}] = \Omega[\bar{H}]$ and $[\ddot{H}] = \Omega^2[\bar{\bar{H}}]$, the temporal derivatives of the governing matrix are

$$[\dot{M}] = \Omega \Big[[0] \quad [\bar{B}_0] \quad [\bar{H}] \quad [\bar{H}][A] \quad [\bar{H}] \Big]$$

and

$$[\ddot{M}] = \Omega^2 \Big[[0] \quad [\bar{\bar{B}}_0] \quad [\bar{\bar{H}}] \quad [\bar{\bar{H}}][A] \quad [\bar{\bar{H}}] \Big].$$

Finally, the three matrices participating in the Lagrange equation for this scenario become

$$[M]^T[M] = \begin{bmatrix} [I] \\ [B_0]^T \\ [H]^T \\ [A]^T[H]^T \\ [H]^T \end{bmatrix} \Big[[I] \quad [B_0] \quad [H] \quad [H][A] \quad [H] \Big]$$

$$= \begin{bmatrix} [I] & [B_0] & [H] & [H][A] & [H] \\ [B_0]^T & [B_0]^T[B_0] & [B_0]^T[H] & [B_0]^T[H][A] & [B_0]^T[H] \\ [H]^T & [H]^T[B_0] & [H]^T[H] & [H]^T[H][A] & [H]^T[H] \\ [A]^T[H]^T & [A]^T[H]^T[B_0] & [A]^T[H]^T[H] & [A]^T[H]^T[H][A] & [A]^T[H]^T[H] \\ [H]^T & [H]^T[B_0] & [H]^T[H] & [H]^T[H][A] & [H]^T \end{bmatrix},$$

$$[M]^T[\dot{M}] = \Omega \begin{bmatrix} [I] \\ [B_0]^T \\ [H]^T \\ [A]^T[H]^T \\ [H]^T \end{bmatrix} \Big[[0] \quad [\bar{B}_0] \quad [\bar{H}] \quad [\bar{H}][A] \quad [\bar{H}] \Big]$$

$$= \begin{bmatrix} [0] & [\bar{B}_0] & [\bar{H}] & [\bar{H}][A] & [\bar{H}] \\ [0] & [B_0]^T[\bar{B}_0] & [B_0]^T[\bar{H}] & [B_0]^T[\bar{H}][A] & [B_0]^T[\bar{H}] \\ [0] & [H]^T[\bar{B}_0] & [H]^T[\bar{H}] & [H]^T[\bar{H}][A] & [H]^T[\bar{H}] \\ [0] & [A]^T[H]^T[\bar{B}_0] & [A]^T[H]^T[\bar{H}] & [A]^T[H]^T[\bar{H}][A] & [A]^T[H]^T[\bar{H}] \\ [0] & [H]^T[\bar{B}_0] & [H]^T[\bar{H}] & [H]^T[\bar{H}][A] & [H]^T[\bar{H}] \end{bmatrix},$$

and

$$[M]^T[\ddot{M}] = \Omega^2 \begin{bmatrix} [I] \\ [B_0]^T \\ [H]^T \\ [A]^T[H]^T \\ [H]^T \end{bmatrix} \begin{bmatrix} [0] & [\bar{\bar{B}}_0] & [\bar{\bar{H}}] & [\bar{\bar{H}}][A] & [\bar{\bar{H}}] \end{bmatrix}$$

$$= \begin{bmatrix} [0] & [\bar{\bar{B}}_0] & [\bar{\bar{H}}] & [\bar{\bar{H}}][A] & [\bar{\bar{H}}] \\ [0] & [B_0]^T[\bar{\bar{B}}_0] & [B_0]^T[\bar{\bar{H}}] & [B_0]^T[\bar{\bar{H}}][A] & [B_0]^T[\bar{\bar{H}}] \\ [0] & [H]^T[\bar{\bar{B}}_0] & [H]^T[\bar{\bar{H}}] & [H]^T[\bar{\bar{H}}][A] & [H]^T[\bar{\bar{H}}] \\ [0] & [A]^T[H]^T[\bar{\bar{B}}_0] & [A]^T[H]^T[\bar{\bar{H}}] & [A]^T[H]^T[\bar{\bar{H}}][A] & [A]^T[H]^T[\bar{\bar{H}}] \\ [0] & [H]^T[\bar{\bar{B}}_0] & [H]^T[\bar{\bar{H}}] & [H]^T[\bar{\bar{H}}][A] & [H]^T[\bar{\bar{H}}] \end{bmatrix}.$$

The matrices partitioned and assembled into the complete Lagrange equation of the fully coupled system considering all four generalized coordinates are of the following form:

$$m \begin{bmatrix} [I] & [B_0] & [H] & [H][A] \\ [B_0]^T & [B_0]^T[B_0] & [B_0]^T[H] & [B_0]^T[H][A] \\ [H]^T & [H]^T[B_0] & [I] & [A] \\ [A]^T[H]^T & [A]^T[H]^T[B_0] & [A]^T & [A]^T[A] \end{bmatrix} \begin{Bmatrix} \ddot{\sigma} \\ \ddot{\beta} \\ \ddot{\rho} \\ \ddot{\alpha} \end{Bmatrix}$$

$$+ 2m\Omega \begin{bmatrix} 0 & [\bar{B}_0] & [\bar{H}] & [\bar{H}][A] \\ 0 & [B_0]^T[\bar{B}_0] & [B_0]^T[\bar{H}] & [B_0]^T[\bar{H}][A] \\ 0 & [H]^T[\bar{B}_0] & [P]^T & [P]^T[A] \\ 0 & [A]^T[H]^T[\bar{B}_0] & [A]^T[P]^T & [A]^T[P]^T[A] \end{bmatrix} \begin{Bmatrix} \dot{\sigma} \\ \dot{\beta} \\ \dot{\rho} \\ \dot{\alpha} \end{Bmatrix}$$

$$+ m\Omega^2 \begin{bmatrix} 0 & [\bar{\bar{B}}_0] & [\bar{\bar{H}}] & [\bar{\bar{H}}][A] \\ 0 & [B_0]^T[\bar{\bar{B}}_0] & [B_0]^T[\bar{\bar{H}}] & [B_0]^T[\bar{\bar{H}}][A] \\ 0 & [H]^T[\bar{\bar{B}}_0] & -[J] & -[J][A] \\ 0 & [A]^T[H]^T[\bar{\bar{B}}_0] & -[A]^T[J] & -[A]^T[J][A] \end{bmatrix} \begin{Bmatrix} \sigma \\ \beta \\ \rho \\ \alpha \end{Bmatrix}$$

$$= m\Omega^2 \begin{bmatrix} -[\bar{\bar{H}}] \\ -[B_0]^T[\bar{\bar{H}}] \\ [J] \\ [A]^T[J] \end{bmatrix} \{r\}. \tag{2.23}$$

In Equation 2.23 the products $[H]^T[H]=[I]$; $[H]^T[\bar{H}]=[P]^T$; $[H]^T[\bar{\bar{H}}]=-[J]$ introduced earlier were used. Following the previous sections' matrices, this most complete coupled equation becomes

$$
\begin{bmatrix}
[M_\sigma] & m[B_0] & m[H] & m[H][A] \\
m[B_0]^T & m[B_0]^T[B_0] & m[B_0]^T[H] & m[B_0]^T[H][A] \\
m[H]^T & m[H]^T[B_0] & [M_\rho] & m[A] \\
m[A]^T[H]^T & m[A]^T[H]^T[B_0] & m[A]^T & [M_\alpha]
\end{bmatrix}
\begin{Bmatrix}
\ddot{\sigma} \\
\ddot{\beta} \\
\ddot{\rho} \\
\ddot{\alpha}
\end{Bmatrix}
$$

$$
+ 2\Omega
\begin{bmatrix}
0 & m[\bar{B}_0] & m[\bar{H}] & m[\bar{H}][A] \\
0 & m[B_0]^T[\bar{B}_0] & m[B_0]^T[\bar{H}] & m[B_0]^T[\bar{H}][A] \\
0 & m[H]^T[\bar{B}_0] & [C_\rho] & m[P]^T[A] \\
0 & m[A]^T[H]^T[\bar{B}_0] & m[A]^T[P]^T & [C_\alpha]
\end{bmatrix}
\begin{Bmatrix}
\dot{\sigma} \\
\dot{\beta} \\
\dot{\rho} \\
\dot{\alpha}
\end{Bmatrix}
$$

$$
+ \Omega^2
\begin{bmatrix}
0 & m[\bar{\bar{B}}_0] & m[\bar{\bar{H}}] & m[\bar{\bar{H}}][A] \\
0 & m[B_0]^T[\bar{\bar{B}}_0] & m[B_0]^T[\bar{\bar{H}}] & m[B_0]^T[\bar{\bar{H}}][A] \\
0 & m[H]^T[\bar{\bar{B}}_0] & [Z_\rho] & -m[J][A] \\
0 & m[A]^T[H]^T[\bar{\bar{B}}_0] & -m[A]^T[J] & [Z_\alpha]
\end{bmatrix}
\begin{Bmatrix}
\sigma \\
\beta \\
\rho \\
\alpha
\end{Bmatrix}
$$

$$
= \Omega^2
\begin{bmatrix}
-m[\bar{\bar{H}}]\{r\} \\
-m[B_0]^T[\bar{\bar{H}}]\{r\} \\
\{f_{c\rho}\} \\
\{f_{c\alpha}\}
\end{bmatrix}.
\tag{2.24}
$$

It is interesting to notice the gradual progression of the equations of the last three sections. They reveal how the computational formulation reflects the underlying increasing complexity of the physical phenomenon, and the rows of this equation also express a very orderly structure. Equation (2.24) is the same as Equation (2.14) after removing the second row and the second column that are related to the stationary part rotation. Similarly, removing the first row and the column of Equation (2.14) results in Equation (2.9).

Furthermore, certain coupling terms related to the [A] matrix may be simplified or even eliminated depending on the nodal mass and inertia definition. The details of these are left to the reader; we leave with the generic formulations presented in Equation (2.23). This equation may also be gradually simplified to return to the intermediate scenarios presented in the previous sections.

This, as well as the other equilibrium equations presented in the prior sections, contains time-dependent terms describing the periodic nature of the rotational phenomenon. In the following section we detail the computational forms of some of those terms. A similar approach was used in ref. [32].

2.5 Time-Dependent Terms of Equations

Most of the terms in Equation (2.23) are time dependent and implicitly periodic. For discussion we will explore the second row of Equation (2.23) introduced in the full coupling scenario in Section 2.4. The $[B_0]$ matrix prominently occurring in this row may be simply partitioned to clearly reveal its periodic nature explicitly as

$$[B_0] = \begin{bmatrix} 0 & 0 & -x\sin\Omega t - y\cos\Omega t \\ 0 & 0 & x\cos\Omega t - y\sin\Omega t \\ x\sin\Omega t + y\cos\Omega t & -x\cos\Omega t + y\sin\Omega t & 0 \end{bmatrix}$$

$$= \begin{bmatrix} 0 & 0 & -x \\ 0 & 0 & -y \\ x & y & 0 \end{bmatrix} \sin\Omega t + \begin{bmatrix} 0 & 0 & -y \\ 0 & 0 & x \\ y & -x & 0 \end{bmatrix} \cos\Omega t = [B_1]\sin\Omega t + [B_2]\cos\Omega t.$$

$$(2.25)$$

The new matrices are introduced for computational convenience and to represent the periodic nature.

The second term in the second row of Equation (2.23) contains the product $[B_0]^T [B_0]$, which gives rise to periodic terms with double rotor speed. Using the relations developed above,

$$[B_0]^T[B_0] = ([B_1]^T\sin\Omega t + [B_2]^T\cos\Omega t)([B_1]\sin\Omega t + [B_2]\cos\Omega t)$$

$$= [B_1]^T[B_1]\sin^2\Omega t + \tfrac{1}{2}([B_1]^T[B_2] + [B_2]^T[B_1])\sin 2\Omega t + [B_2]^T[B_2]\cos^2\Omega t.$$

$$(2.26)$$

The above equation may be brought to the final form of

$$[B_0]^T[B_0] = [B_3] + [B_4]\sin 2\Omega t + [B_5]\cos 2\Omega t.$$

$$(2.27)$$

The content of the newly introduced matrices may easily be obtained by simple substitution and employing known trigonometric identities. The third term in the same second row in Equation 2.23 is the $[B_0]^T[H]$ product. The product in detail is

$$[B_0]^T[H] = ([B_1]^T \sin \Omega t + [B_2]^T \cos \Omega t)[H] = [B_1]^T[H]\sin \Omega t + [B_2]^T[H]\cos \Omega t$$

$$= \begin{bmatrix} 0 & 0 & x \\ 0 & 0 & y \\ -x & -y & 0 \end{bmatrix} \begin{bmatrix} \cos \Omega t & -\sin \Omega t & 0 \\ \sin \Omega t & \cos \Omega t & 0 \\ 0 & 0 & 1 \end{bmatrix} \sin \Omega t$$

$$+ \begin{bmatrix} 0 & 0 & y \\ 0 & 0 & -x \\ -y & x & 0 \end{bmatrix} \begin{bmatrix} \cos \Omega t & -\sin \Omega t & 0 \\ \sin \Omega t & \cos \Omega t & 0 \\ 0 & 0 & 1 \end{bmatrix} \cos \Omega t$$

$$= \begin{bmatrix} 0 & 0 & x \\ 0 & 0 & y \\ -x & -y & 0 \end{bmatrix} \begin{bmatrix} \sin \Omega t \cos \Omega t & -\sin^2 \Omega t & 0 \\ \sin^2 \Omega t & \sin \Omega t \cos \Omega t & 0 \\ 0 & 0 & \sin \Omega t \end{bmatrix}$$

$$+ \begin{bmatrix} 0 & 0 & y \\ 0 & 0 & -x \\ -y & x & 0 \end{bmatrix} \begin{bmatrix} \cos^2 \Omega t & -\sin \Omega t \cos \Omega t & 0 \\ \sin \Omega t \cos \Omega t & \cos^2 \Omega t & 0 \\ 0 & 0 & \cos \Omega t \end{bmatrix}$$

$$= \begin{bmatrix} 0 & 0 & x \sin \Omega t \\ 0 & 0 & y \sin \Omega t \\ -x \sin \Omega t \cos \Omega t - y \sin^2 \Omega t & x \sin^2 \Omega t - y \sin \Omega t \cos \Omega t & 0 \end{bmatrix}$$

$$+ \begin{bmatrix} 0 & 0 & y \cos \Omega t \\ 0 & 0 & -x \cos \Omega t \\ x \sin \Omega t \cos \Omega t - y \cos^2 \Omega t & x \cos^2 \Omega t + y \sin \Omega t \cos \Omega t & 0 \end{bmatrix}$$

$$= \begin{bmatrix} 0 & 0 & x \sin \Omega t + y \cos \Omega t \\ 0 & 0 & y \cos \Omega t - x \sin \Omega t \\ -y & x & 0 \end{bmatrix}.$$

$$(2.28)$$

The bottom row loses its time dependency and becomes a constant due to cancellations. The last term in the second row of Equation (2.23) is the product

$$[B_0]^T[H][A].$$

Let us view the transpose $[A]^T[H]^T$ of part of the component and use a fictitious vector to multiply this product. The following reformulation of the term reveals its periodic nature as

$$
[A]^T[H]^T\{v\} =
\begin{bmatrix} 0 & -z' & y' \\ z' & 0 & -x' \\ -y' & x' & 0 \end{bmatrix}
\begin{bmatrix} \cos\Omega t & \sin\Omega t & 0 \\ -\sin\Omega t & \cos\Omega t & 0 \\ 0 & 0 & 1 \end{bmatrix}
\begin{Bmatrix} a \\ b \\ c \end{Bmatrix}
$$

$$
=
\begin{bmatrix} 0 & -z' & y' \\ z' & 0 & -x' \\ -y' & x' & 0 \end{bmatrix}
\begin{Bmatrix} a\cos\Omega t + b\sin\Omega t \\ -a\sin\Omega t + b\cos\Omega t \\ c \end{Bmatrix}
$$

$$
=
\begin{Bmatrix} z'(a\sin\Omega t - b\cos\Omega t) + y'c \\ z'(a\cos\Omega t + b\sin\Omega t) - x'c \\ -y'(a\cos\Omega t + b\sin\Omega t) + x'(-a\sin\Omega t + b\cos\Omega t) \end{Bmatrix}
$$

$$
= \sin\Omega t
\begin{Bmatrix} z'a \\ z'b \\ -y'b - x'a \end{Bmatrix}
+ \cos\Omega t
\begin{Bmatrix} -z'b \\ z'a \\ -y'a + x'b \end{Bmatrix}
+
\begin{Bmatrix} y'c \\ -x'c \\ 0 \end{Bmatrix}
$$

$$
= \sin\Omega t
\begin{bmatrix} z' & 0 & 0 \\ 0 & z' & 0 \\ -x' & -y' & 0 \end{bmatrix}
\begin{Bmatrix} a \\ b \\ c \end{Bmatrix}
+ \cos\Omega t
\begin{bmatrix} 0 & -z' & 0 \\ z' & 0 & 0 \\ -y' & x' & 0 \end{bmatrix}
\begin{Bmatrix} a \\ b \\ c \end{Bmatrix}
$$

$$
+
\begin{bmatrix} 0 & 0 & y' \\ 0 & 0 & -x' \\ 0 & 0 & 0 \end{bmatrix}
\begin{Bmatrix} a \\ b \\ c \end{Bmatrix}.
$$

$$(2.29)$$

Introducing new matrices, we can write

$$[A]^T[H]^T\{v\} = ([A_0] + [A_1]\sin\Omega t + [A_2]\cos\Omega t)\{v\}.$$

Hence the quantity we sought becomes

$$[H][A] = [A_0]^T + [A_1]^T\sin\Omega t + [A_2]^T\cos\Omega t.$$

$$(2.30)$$

This now is premultiplied to produce the component

$$[B_0]^T [H][A] = ([B_1]^T \sin \Omega t + [B_2]^T \cos \Omega t)([A_0]^T + [A_1]^T \sin \Omega t + [A_2]^T \cos \Omega t)$$

$$= [B_1]^T [A_0]^T \sin \Omega t + [B_2]^T [A_0]^T \cos \Omega t + [B_1]^T [A_1]^T \sin^2 \Omega t$$

$$+ [B_2]^T [A_2]^T \cos^2 \Omega t + \tfrac{1}{2}([B_1]^T [A_2]^T + [B_2]^T [A_1]^T) \sin 2\Omega t.$$

(2.31)

We again obtain double periods. Evaluating all the individual matrix products and employing some trigonometric identities, the expression may be further simplified.

Finally, on the right-hand side, the centrifugal term in the second row of Equation (2.23) contains the product $[\bar{H}]\{r\}$, which may also be restructured as

$$[\bar{H}]\{r\} = \begin{bmatrix} -\cos \Omega t & \sin \Omega t & 0 \\ -\sin \Omega t & -\cos \Omega t & 0 \\ 0 & 0 & 0 \end{bmatrix} \begin{Bmatrix} x \\ y \\ z \end{Bmatrix} = \begin{Bmatrix} -x \cos \Omega t + y \sin \Omega t \\ -x \sin \Omega t - y \cos \Omega t \\ 0 \end{Bmatrix}$$

$$= \sin \Omega t \begin{Bmatrix} y \\ -x \\ 0 \end{Bmatrix} + \cos \Omega t \begin{Bmatrix} -x \\ -y \\ 0 \end{Bmatrix}.$$

The second right-hand-side term then shows its periodic nature in the form of

$$[B_0]^T [H]\{r\} = \left([B_1]^T \sin \Omega t + [B_2]^T \cos \Omega t)(\sin \Omega t \begin{Bmatrix} y \\ -x \\ 0 \end{Bmatrix} + \cos \Omega t \begin{Bmatrix} -x \\ -y \\ 0 \end{Bmatrix} \right)$$

$$= \tfrac{1}{2}([B_2]^T \begin{Bmatrix} y \\ -x \\ 0 \end{Bmatrix} + [B_1]^T \begin{Bmatrix} -x \\ -y \\ 0 \end{Bmatrix}) \sin 2\Omega t = \begin{Bmatrix} 0 \\ 0 \\ x^2 + y^2 \end{Bmatrix} \sin 2\Omega t.$$

(2.32)

The coefficients of the squares of the trigonometric functions became zero in the above equation, and the result is a force proportional to the particle's radial distance from the axis of rotation. This concludes all the terms in the second row of Equation (2.23). The terms in the other rows and in the other block matrices are similar in structure, except that some of the participating matrices are replaced by their derivatives. These will introduce some sign changes, but similar formulae will apply. The details of these are not presented as they do not carry any more instructional value.

In conclusion, after introducing appropriate submatrices for all the time-dependent terms, the final equation of motion for the fully coupled scenario becomes

$$[M]\{\ddot{g}\} + \Omega[C]\{\dot{g}\} - \Omega^2[Z]\{g\} = \{f\}, \tag{2.33}$$

where the system matrices are a collection of periodic component matrices in the form of

$$[M] = [M_0] + \sin \Omega t[M_1] + \cos \Omega t[M_2] + \sin 2\Omega t[M_3] + \cos 2\Omega t[M_4], \tag{2.34}$$

$$[C] = [C_0] + \sin \Omega t[C_1] + \cos \Omega t[C_2] + \sin 2\Omega t[C_3] + \cos 2\Omega t[C_4], \tag{2.35}$$

and

$$[Z] = [Z_0] + \sin \Omega t[Z_1] + \cos \Omega t[Z_2] + \sin 2\Omega t[Z_3] + \cos 2\Omega t[Z_4]. \tag{2.36}$$

The right-hand-side force vector also exhibits the behavior

$$\{f\} = \{f_0\} + \{f_1\} \sin \Omega t + \{f_2\} \cos \Omega t + \{f_3\} \sin 2\Omega t. \tag{2.37}$$

Further details of this submatrix partitioning may be important in software implementation, but, as such, is beyond our focus. The salient point is in the recognition of the periodic nature of these components and their effect on certain application scenarios, as well as on the numerical solution technology. The latter topic is addressed in Chapter 5, Section 5.4 and the applications are the focus of Part II.

3

Finite Element Analysis of
Rotating Structures

A rotating structure usually consists of many particles, and the assembly of the equations of motions for all particles results in the equilibrium of the complete mechanical system. In the case of a structure with many (for example, n) particles, each will have generalized coordinates g_i containing the nodal translational (u_i, v_i, w_i) and rotational $(\varphi_i, \psi_i, \theta_i)$ displacements:

$$\{g\} = \begin{Bmatrix} g_1 \\ g_2 \\ \cdots \\ g_i \\ \cdots \\ g_n \end{Bmatrix} \qquad \{g_i\} = \begin{Bmatrix} u_i \\ v_i \\ w_i \\ \varphi_i \\ \psi_i \\ \theta_i \end{Bmatrix}. \tag{3.1}$$

The order of the nodal displacement components is chosen to be translations followed by rotations, but it is not a necessity. Most commercial software use this order, but the choice has no bearing on the following discussions.

The equilibrium of a mechanical system is a system of the complete (sometimes called second type) Lagrange equations of motion of the form

$$\frac{d}{dt}\left(\frac{\partial L}{\partial \{\dot{g}_i\}} \right) - \frac{\partial L}{\partial \{g_i\}} + \frac{dD}{d\{\dot{g}_i\}} = \frac{dW}{d\{g_i\}} ; i = 1,2,...n. \tag{3.2}$$

In the equations L is the Lagrange potential, also known as the Lagrangian. It is the difference of the kinetic and potential energy of the particle,

$$L = T - U. \tag{3.3}$$

D is the dissipative energy, and W is the work of nondissipative forces. The potential energy of the particle is not dependent on time; hence its contribution to the first term of the equation is zero. Therefore, the more usual and intuitive form of the equation is

$$\frac{d}{dt}\left(\frac{\partial T}{\partial \{\dot{g}_i\}} \right) - \frac{\partial T}{\partial \{g_i\}} + \frac{\partial U}{\partial \{g_i\}} + \frac{dD}{d\{\dot{g}_i\}} = \frac{dW}{d\{g_i\}} ; i = 1,2,...n. \tag{3.4}$$

This equation justifies the foundation we laid in the prior two chapters, focusing on the kinetic energy only, because all the equations derived are directly applicable with appropriate augmentation with other terms. The left-hand side contains another velocity-dependent term, reflecting the dissipative forces in the system, and the right-hand side of the equation contains the nondissipative forces' contribution to the equilibrium. They are discussed in the following sections after the potential energy contribution.

3.1 Potential Energy of Structure

The rotating particle is part of a structural system that undergoes elastic deformations. The potential energy of the structure due to elastic deformations is

$$U = \frac{1}{2}\int (\{\sigma\}_0 + \{\sigma\})^T (\{\varepsilon\} + \{\varepsilon\}_l)\, dV$$

$$= \frac{1}{2}\int \left(\{\sigma\}_0^T \{\varepsilon\} + \{\sigma\}_0^T \{\varepsilon\}_l + \{\sigma\}^T \{\varepsilon\} + \{\sigma\}^T \{\varepsilon\}_l \right) dV. \tag{3.5}$$

Here the integral is taken over the volume of the structure, and $\{\sigma\}$, $\{\varepsilon\}$ are the linear stresses and strains of the structure, respectively.[*] The $\{\sigma\}_0$ and $\{\varepsilon\}_l$ vectors represent the initial stresses and the large strains, respectively. For this discussion we'll ignore the first and last products as second-order small quantities. The remaining terms are separated into two integrals as

$$U = \frac{1}{2}\int \left(\{\sigma\}_0^T \{\varepsilon\}_l \right) dV + \frac{1}{2}\int \{\sigma\}^T \{\varepsilon\}\, dV. \tag{3.6}$$

We introduce the linear stress-strain relationship $\{\sigma\} = [E]\{\varepsilon\}$, where

$$\{\sigma\} = \begin{Bmatrix} \sigma_{xx} \\ \sigma_{yy} \\ \sigma_{zz} \\ \sigma_{xy} \\ \sigma_{xz} \\ \sigma_{yz} \end{Bmatrix} \quad \text{and} \quad \{\varepsilon\} = \begin{Bmatrix} \partial u/\partial x \\ \partial v/\partial y \\ \partial w/\partial z \\ \partial u/\partial y + \partial v/\partial x \\ \partial u/\partial z + \partial w/\partial x \\ \partial v/\partial z + \partial w/\partial x \end{Bmatrix}. \tag{3.7}$$

---

[*] Note that this is not the same $\{\sigma\}$ vector introduced in Chapter 1, Equation 1.1. However, we use the same letter here to adhere to industry convention.

The $[E]$ is the material matrix containing the Young's modulus and the Poisson ratio.

The second integral of Equation (3.6) represents the elastic potential energy of the structure:

$$U_E = \frac{1}{2} \int \{\sigma\}^T \{\varepsilon\} \, dV = \frac{1}{2} \int \{\varepsilon\}^T [E] \{\varepsilon\} \, dV. \tag{3.8}$$

The linear strain is a function of the nodal generalized displacements as

$$\{\varepsilon\} = [B]\{g\}, \tag{3.9}$$

where $[B]$ is the matrix of shape functions dependent on the type of elements (solid or shell) used in modeling.* With the newly introduced quantities, the elastic potential energy is

$$U_E = \frac{1}{2} \int \{g\}^T [B]^T [E][B] \{g\} \, dV. \tag{3.10}$$

Considering our aim of the complete Lagrange equation of motion, the derivative of the elastic potential energy component, after cancellation by 2, becomes

$$\frac{dU_E}{d\{g\}} = \int [B]^T [E][B] dV \, \{g\} = [K]\{g\}. \tag{3.11}$$

This yields the conventional finite element stiffness matrix of the structure:

The large strain vector from the translations of any point $\{\rho\} = [u \quad v \quad w]^T$ of the structure is

$$\{\varepsilon\}_l = \left\{ \begin{array}{c} \{\rho\}_x^T \{\rho\}_x \\ \{\rho\}_y^T \{\rho\}_y \\ \{\rho\}_z^T \{\rho\}_z \\ \{\rho\}_x^T \{\rho\}_y + \{\rho\}_y^T \{\rho\}_x \\ \{\rho\}_x^T \{\rho\}_z + \{\rho\}_z^T \{\rho\}_x \\ \{\rho\}_y^T \{\rho\}_z + \{\rho\}_z^T \{\rho\}_y \end{array} \right\}.$$

where $\{\rho\}_* = \partial\{\rho\}/\partial*; * = x, y, z$.

---

* Note that his is not the same B matrix introduced in Chapter 2, Equation 2.16. However, we use the same letter here to adhere to industry convention

The first integral in Equation (3.6) represents the potential energy of the structure due to the initial stress. Exploiting that $\sigma_{**}\{\rho\}_*^T \{\rho\}_* = \{\rho\}_*^T \sigma_{**}\{\rho\}_*$ for all terms and rearranging results in

$$U_G = \frac{1}{2}\int \left(\{\sigma\}_0^T \{\varepsilon\}_l\right) dV = \int \{e\}_l^T [S]_0 \{e\}_l \, dV. \tag{3.12}$$

The middle matrix contains the initial stress terms in the form of

$$[S]_0 = \begin{bmatrix} I_3\sigma_{xx} & I_3\sigma_{xy} & I_3\sigma_{xz} \\ I_3\sigma_{yx} & I_3\sigma_{yy} & I_3\sigma_{yx} \\ I_3\sigma_{zx} & I_3\sigma_{zy} & I_3\sigma_{zz} \end{bmatrix} ; I_3 = \begin{bmatrix} 1 & 0 & 0 \\ 0 & 1 & 0 \\ 0 & 0 & 1 \end{bmatrix}. \tag{3.13}$$

Without detail, the $\{e\}_l$ vector is related to the displacements via the derivatives of the shape functions as $\{e\}_l = [B']\{g\}$. The initial stress is the result of external forces (for example, the centrifugal force) and is dependent on the rotation speed. Hence the derivative of the initial stress-based potential energy yields

$$\frac{dU_G}{d\{g\}} = \int [B']^T [S]_0 [B'] \, dV \{g\} = \Omega^2 [K_G]\{g\}. \tag{3.14}$$

This is the geometric or differential stiffness matrix whose size is the same as the elastic stiffness matrix, namely, six times the number of node points. For more details see ref. [1].

3.2 Dissipative Forces

The next term on the left-hand side of the general Lagrange equation is related to the dissipative forces acting on the structure representing the energy lost in the structure. Those forces are damping the kinetic energy of the particles; hence they will produce the damping matrices. The dissipative forces may be internal or external.

The internal dissipative function represents the damping between the particles of the rotor and is of the form of the so-called Rayleigh function

$$D_f = \frac{1}{2}\sum_{i=1}^{n}\sum_{j=1}^{n} d_{i,j}\{\dot{g}_i\}^T \{\dot{g}_j\}. \tag{3.15}$$

Here the coefficients define the internal damping exerted between the generalized coordinates related to two node points: i and j. Let us focus on a particular node point and its translational nodal displacement in the fixed system. The dissipative function at that node point, as a result of the other points, is

$$D_{\rho,i} = \frac{1}{2} d_{i,i} \{\dot{\rho}\}_i^T \{\dot{\rho}\}_i. \tag{3.16}$$

Following Chapter Equations (1.30) and (1.31), we may write

$$\{\dot{\rho}\} = [\dot{H}]\{\rho\} + [H]\{\dot{\rho}\}. \tag{3.17}$$

Hence the dissipative function of the point in the rotational coordinate system will be

$$D_{\rho,i} = \frac{1}{2} d_{i,i} ([\dot{H}]\{\rho\}_i + [H]\{\dot{\rho}\}_i)^T ([\dot{H}]\{\rho\}_i + [H]\{\dot{\rho}\}_i). \tag{3.18}$$

Executing the posted operations yields

$$D_{\rho,i} = \frac{1}{2} d_{i,i} \left(\{\rho\}_i^T [\dot{H}]^T [\dot{H}]\{\rho\}_i + \{\rho\}_i^T [\dot{H}]^T [H]\{\dot{\rho}\}_i + \{\dot{\rho}\}_i^T [H]^T [\dot{H}]\{\rho\}_i \right.$$
$$\left. + \{\dot{\rho}\}_i^T [H]^T [H]\{\dot{\rho}\}_i \right), \tag{3.19}$$

where the middle two terms cancel each other out because $[H]^T[\dot{H}] = -[\dot{H}]^T[H]$. Furthermore, the first term is not dependent on the velocity; hence the contribution of the dissipative function to Lagrange's equation is

$$\frac{dD_{\rho,i}}{d\{\dot{\rho}\}_i} = [D_\rho]_i \{\dot{\rho}\}_i. \tag{3.20}$$

The matrix is formed as

$$[D_\rho]_i = d_{i,i}[I] = \begin{bmatrix} d_{i,i} & & \\ & d_{i,i} & \\ & & d_{i,i} \end{bmatrix}. \tag{3.21}$$

Similar considerations for the rotational nodal displacements result in a matrix of

$$[D_\alpha]_i = d_{i,i}[I] = \begin{bmatrix} d_{i,i} & & \\ & d_{i,i} & \\ & & d_{i,i} \end{bmatrix}. \qquad (3.22)$$

The augmentation of the two matrices will represent the internal damping forces in the finite element equation of motion

$$[D_l]_i = \begin{bmatrix} d_{i,u} & & & & & \\ & d_{i,v} & & & & \\ & & d_{i,w} & & & \\ & & & d_{i,\varphi} & & \\ & & & & d_{i,\psi} & \\ & & & & & d_{i,\theta} \end{bmatrix}, \qquad (3.23)$$

where the subscript I represents the internal nature of this damping. The indices now represent the possibility that the damping force components acting on the i-th mass particle may be different with respect to the rotational and translational nodal displacements.

The external dissipative forces are usually originated in the bearing and will be the subject of a more detailed discussion in Chapter 8, Section 8.4. These forces carry the effect of the bearing to the rotating part of the structure, mainly but not necessarily only in a plane perpendicular to the axis of rotation. These forces are proportional to the nodal velocity of the point. They are given in the fixed coordinate system that governs the stationary part as

$$\{F\}_i = \begin{Bmatrix} \bar{F}_x \\ \bar{F}_y \\ 0 \end{Bmatrix}_i = -d_{F,i} \begin{Bmatrix} \dot{u} \\ \dot{v} \\ 0 \end{Bmatrix}_i = -d_{F,i}\{\dot{\rho}\}_i. \qquad (3.24)$$

The force acting on the rotating part in the rotating coordinate system is related as

$$\{\bar{F}\} = H\{F\}. \qquad (3.25)$$

Because the relationship between vectors in the two systems is $\{\bar{r}\} = [H]\{r\}$, the velocities are also related likewise, as shown in Equation (3.17):

$$\{\dot{\bar{\rho}}\} = \Omega[\bar{H}]\{\rho\}_i + [H]\{\dot{\rho}\}_i. \tag{3.26}$$

Hence the force in the fixed system is

$$\{\bar{F}\}_i = -d_{F,i}\Omega[\bar{H}]\{\rho\}_i - d_{F,i}[H]\{\dot{\rho}\}_i. \tag{3.27}$$

Because $[H]^T[H] = [I]$, it follows that $\{F\}_i = [H]^T\{\bar{F}\}_i$. Substituting yields the damping force in the rotating system as

$$\{F\}_i = -d_{F,i}\Omega[H]^T[\bar{H}]\{\rho\}_i - d_{F,i}[H]^T[H]\{\dot{\rho}\}_i. \tag{3.28}$$

Finally, using the earlier matrix relations, the bearing-induced damping force in the rotating reference system becomes

$$\{F\}_i = -\Omega d_{F,i}[P]^T\{\rho\}_i - d_{F,i}[I]\{\dot{\rho}\}_i. \tag{3.29}$$

These forces are functions of the displacement of the adjacent particles and their velocity, as we can now see. The first term, proportional to the displacement, is called the circulatory matrix. It is an antisymmetric matrix by virtue of its defining matrix as

$$d_{F,i}[P]^T = \begin{bmatrix} 0 & d_{F,i} & 0 \\ -d_{F,i} & 0 & 0 \\ 0 & 0 & 0 \end{bmatrix} = [K_D]_{\rho,i}. \tag{3.30}$$

Note that the $[K_D]$ name is chosen to reflect the fact that it is a damping component associated with the stiffness term. The second term simply yields

$$d_{F,i}[I] = \begin{bmatrix} d_{F,i} & & \\ & d_{F,i} & \\ & & d_{F,i} \end{bmatrix} = [D_E]_{\rho,i}. \tag{3.31}$$

In certain types of bearings, there are also moments conveyed to the rotor by the bearing. These result in a relationship with the nodal rotations:

$$\{\bar{M}\}_i = \left\{ \begin{array}{c} \bar{M}_x \\ \bar{M}_y \\ 0 \end{array} \right\}_i = -d_{M,i} \left\{ \begin{array}{c} \dot{\bar{\varphi}} \\ \dot{\bar{\psi}} \\ 0 \end{array} \right\}_i = -d_{M,i} \{\dot{\bar{\alpha}}\}_i. \tag{3.32}$$

Having identical steps as above and assembling the translational and rotational partitions result in the finite element matrices for the i-th particle being

$$[K_D]_i = \begin{bmatrix} 0 & d_{F,i} & 0 & & & \\ -d_{F,i} & 0 & 0 & & & \\ 0 & 0 & 0 & & & \\ & & & 0 & d_{M,i} & 0 \\ & & & -d_{M,i} & 0 & 0 \\ & & & 0 & 0 & 0 \end{bmatrix}. \tag{3.33}$$

Similarly, the two second components produce the external damping matrix

$$[D_E]_i = \begin{bmatrix} d_{F,i} & & & & & \\ & d_{F,i} & & & & \\ & & d_{F,i} & & & \\ & & & d_{M,i} & & \\ & & & & d_{M,i} & \\ & & & & & d_{M,i} \end{bmatrix}. \tag{3.34}$$

The distinction between the force- and moment-related damping coefficients is encapsulated in the F and M subscripts, and their location indicates a potential difference between components.

The internal and external damping matrices related to the velocity are usually added as $[D]_i = [D_E]_i + [D_I]_i$, because their location and multiplier in the equilibrium equation are identical. This is also called the viscous damping of the system, although not necessarily related to viscous dampers. More complicated bearing models may even bring external loads that are functions of the rotational speed, not just proportional as in the above. This topic is also the subject of further discussion in Chapter 8, Section 8.4.

3.3 Nondissipative Forces

The right-hand-side term of Equation (3.2) represents the nondissipative forces acting on the structure. They are also of two kinds. The first kind constitutes the conservative forces acting on the structure. Such forces may be derived from a vector potential for which the work of the force is independent of the path taken by the body on a closed curve in the vector field. Such forces usually act throughout the body of the structure, hence resulting in a volume integral

$$W_p = -\iiint \{f_p\}\{g\}\, dV. \tag{3.35}$$

An example of such is the force of gravity that obviously acts on each particle of the structural body. For the sake of simplicity, we assume a single vector active force, but that is not necessary. The contribution of the conservative forces to Lagrange's equation is the term

$$\{F_p\} = -\frac{dW_p}{d\{g\}}, \tag{3.36}$$

where $\{F_p\}$ is the global vector of conservative forces acting throughout the volume of the structure.

The second kind of nondissipative forces arises from an active external physical phenomenon interacting with the rotating structure. A structure located in any other medium than a vacuum experiences forces of the fluid field surrounding it. The fluid may be air, as in the case of a wind turbine, or water, as in the case of water turbines. The work exerted on the structure by the external field usually acts on the surface and is defined by a surface integral

$$W_a = -\iint \{f_a\}\{g\}\, dS. \tag{3.37}$$

The result is the term on the right-hand side of Lagrange's equation as

$$\frac{dW_a}{d\{g\}} = \{F_a\}. \tag{3.38}$$

Here $\{F_a\}$ is the global vector of all active forces acting on the surface of the structure. Coupling external forces to the surface of the rotating structure is usually confined to a part of the structure; hence the active surface load vector may be very sparse, containing nonzero components only at selected locations where the external loads connect.

The conservative force vector is denser because it represents something that is uniformly acting on all particles, but sometimes it is not fully populated because the volume force has a certain direction. For example, gravity is unidirectional and may coincide with one of the coordinate axes, so it will have only one component from the six components of each generalized coordinate. The centrifugal force $\{F_c\}$ computed earlier is also a volume force; however, its direction will seldom directly align with any of the coordinate axes as it is periodically changing.

The nondissipative forces may be summed up for all the particles of the structure, and the final vector of forces will be

$$\{F\} = \{F_c\} + \{F_p\} + \{F_a\}. \tag{3.39}$$

This vector will occupy the right-hand side of the equilibrium equation.

The coupling of active external loads will be discussed in detail in Chapter 8, Section 8.5. The interaction between airflow and a rotational structure will be discussed at length in Chapters 9 and 10.

3.4 Finite Element Equation Assembly

The finite element equilibrium equation in the rotating coordinate system becomes

$$[M]\{\ddot{g}\} + ([D] + 2\Omega[C])\{\dot{g}\} + ([K] - \Omega^2[Z] + \Omega^2[K_G] + \Omega[K_D])\{g\} = \{F\}. \tag{3.40}$$

In the fixed coordinate system the finite element equilibrium equation is

$$[M]\{\ddot{g}\} + ([D] + \Omega[C])\{\dot{g}\} + ([K] + \Omega[K_D])\{g\} = \{F\}. \tag{3.41}$$

The visible difference in the latter is the lack of the centrifugal and differential stiffness matrices. The coefficient of the gyroscopic matrix is also changed, as well as its content. Specifically, the Coriolis terms are missing, as they should be, because the force is only detectable in the rotating system.

We first discuss the issue of assembling the rotating nodes of the finite element model. We need to assemble the matrices of each rotating node to the global collection of nodes of the structure. This is a simple mapping as

$$[M] = \begin{bmatrix} [M]_1 & & & \\ & \ddots & & \\ & & [M]_i & \\ & & & \ddots & \\ & & & & [M]_n \end{bmatrix}, \tag{3.42}$$

where all the off-diagonal terms are zero. Note that the mass matrix associated with a rotating node contains the conventional (lumped) mass and the rotational inertia values. Similarly for the gyroscopic and centrifugal matrices, the simple mapping applies:

$$[C] = \begin{bmatrix} [C]_1 & & & \\ & \ddots & & \\ & & [C]_i & \\ & & & \ddots \\ & & & & [C]_n \end{bmatrix}, \qquad (3.43)$$

$$[Z] = \begin{bmatrix} [Z]_1 & & & \\ & \ddots & & \\ & & [Z]_i & \\ & & & \ddots \\ & & & & [Z]_n \end{bmatrix}. \qquad (3.44)$$

The viscous and structural damping matrices are also assembled accordingly as

$$[D] = \begin{bmatrix} [D]_1 & & & \\ & \ddots & & \\ & & [D]_i & \\ & & & \ddots \\ & & & & [D]_n \end{bmatrix}, [K_D] = \begin{bmatrix} [K_D]_1 & & & \\ & \ddots & & \\ & & [K_D]_i & \\ & & & \ddots \\ & & & & [K_D]_n \end{bmatrix}. \qquad (3.45)$$

The finite element matrices now represent the conventional six degrees of freedom, three translations and three rotations for all particles (node points) in the structure. In practical circumstances, however, the structure has a stationary component as well, which is the subject of the next section.

3.5 Coupled Equilibrium Equation Assembly

The finite element equilibrium equation assembly is a bit more difficult when the coupling between the stationary and the rotating parts of the structure is also considered. In the fully coupled scenario presented in Chapter 2,

Section 2.4, the $\{\sigma\}$ offset vector and $\{\beta\}$ rotational vector produce the degrees of freedom of a node of the stationary part:

$$\{g\}_j = \left\{ \begin{array}{c} \{\sigma_j\} \\ \{\beta_j\} \end{array} \right\}. \tag{3.46}$$

The rotating part nodes are represented by

$$\{g\}_i = \left\{ \begin{array}{c} \{\rho_i\} \\ \{\alpha_i\} \end{array} \right\} \tag{3.47}$$

degrees of freedom. Note that because of the presence of two distinct mass particles, we need to distinguish between their masses. The new coupling terms influence the assembly process of the global mass matrix as follows:

$$[M] = \begin{bmatrix} \cdots & & & \\ & [M]_j & [M]_{ji} & \\ & & \cdots & \\ & [M]_{ij} & [M]_i & \\ & & & \cdots \end{bmatrix}, \tag{3.48}$$

where the submatrices are the 2×2 partitions of the coupled mass matrix derived in Chapter 2, Section 2.4. They are

$$[M]_j = m_j \begin{bmatrix} [I] & [B_0] \\ [B_0]^T & [B_0]^T[B_0] \end{bmatrix}, \tag{3.49}$$

$$[M]_{ji} = m_j \begin{bmatrix} [H] & [H][A] \\ [B_0]^T[H] & [B_0]^T[H][A] \end{bmatrix}, \tag{3.50}$$

$$[M]_i = m_i \begin{bmatrix} [I] & [A] \\ [A]^T & [A]^T[A] \end{bmatrix}. \tag{3.51}$$

The coupling of the mass matrix is symmetric because $[M]_{ij} = [M]_{ji}^T$. In the case of the gyroscopic and the centrifugal matrices, this is not the case. The gyroscopic matrix assembly proceeds as

$$[C] = \begin{bmatrix} \ddots & & & \\ & [C]_j & [C]_{ji} & \\ & & \ddots & \\ & [C]_{ij} & [C]_i & \\ & & & \ddots \end{bmatrix}, \tag{3.52}$$

with

$$[C]_j = \begin{bmatrix} [0] & [\bar{B}_0] \\ [0] & [B_0]^T[\bar{B}_0] \end{bmatrix}, \tag{3.53}$$

$$[C]_{ji} = \begin{bmatrix} [\bar{H}] & [\bar{H}][A] \\ [B_0]^T[\bar{H}] & [B_0]^T[\bar{H}][A] \end{bmatrix}, \tag{3.54}$$

and

$$[C]_i = \begin{bmatrix} [P]^T & [P]^T[A] \\ [A]^T[P]^T & [A]^T[P]^T[A] \end{bmatrix}, \tag{3.55}$$

but

$$[C]_{ij} = \begin{bmatrix} [0] & [H]^T[\bar{B}_0] \\ [0] & [A]^T[H]^T[\bar{B}_0] \end{bmatrix} \neq [C]_{ji}^T. \tag{3.56}$$

Similarly, the centrifugal matrices are assembled as

$$[Z] = \begin{bmatrix} \ddots & & & \\ & [Z]_j & [Z]_{ji} & \\ & & \ddots & \\ & [Z]_i & [Z]_i & \\ & & & \ddots \end{bmatrix}, \tag{3.57}$$

with

$$[Z]_j = \begin{bmatrix} [0] & [\bar{\bar{B}}_0] \\ [0] & [B_0]^T[\bar{\bar{B}}_0] \end{bmatrix}, \qquad (3.58)$$

$$[Z]_{ji} = \begin{bmatrix} [\bar{\bar{H}}] & [\bar{\bar{H}}][A] \\ [B_0]^T[\bar{\bar{H}}] & [B_0]^T[\bar{\bar{H}}][A] \end{bmatrix}, \qquad (3.59)$$

$$[Z]_i = \begin{bmatrix} [\bar{\bar{H}}] & [\bar{\bar{H}}][A] \\ [A]^T[\bar{\bar{H}}] & [A]^T[\bar{\bar{H}}][A] \end{bmatrix}, \qquad (3.60)$$

$$[Z]_{ij} = \begin{bmatrix} [0] & [H]^T[\bar{\bar{B}}_0] \\ [0] & [A]^T[H]^T[\bar{\bar{B}}_0] \end{bmatrix}. \qquad (3.61)$$

Again, the coupling is notably unsymmetric. The C and Z matrices are also multiplied by the appropriate mass (m) and rotation speed (Ω) values.

3.6 Analysis Equilibrium Equations

We are now in a position to use the finite element equation of motions of a rotating structure for analysis. First we need to recognize that certain parts do not need to be present in the analysis. Some parts of the model are constrained by boundary conditions restricting their motion. For example, the points of the stationary structure adjacent and attached to the ground may not have nodal displacements at all. This subset of the points is usually constrained with single point constraints; hence they are denoted with the subscript s here. They are simply partitioned out of the global displacement vector as

$$\{g\} = \begin{Bmatrix} \{g_s\} = 0 \\ \{g_k\} \end{Bmatrix}, \qquad (3.62)$$

where the subscript k denotes the remaining points not affected by boundary conditions. The structural matrices participating in the equation of equilibrium may be simply partitioned as

$$[K] = \begin{bmatrix} K_{ss} & K_{sk} \\ K_{ks} & K_{kk} \end{bmatrix}, \tag{3.63}$$

$$[D] = \begin{bmatrix} D_{ss} & D_{sk} \\ D_{ks} & D_{kk} \end{bmatrix}, \tag{3.64}$$

$$[M] = \begin{bmatrix} M_{ss} & M_{sk} \\ M_{ks} & M_{kk} \end{bmatrix}, \tag{3.65}$$

and

$$\{F\} = \begin{Bmatrix} F_s \\ F_k \end{Bmatrix}. \tag{3.66}$$

Here the partition matrix sizes are indicated by their subscripts. These matrices, however, are still not yet ready for analysis.

There may be some components of the structure whose motion is dependent on some other components. For example, rigid components of the structure have a direct relationship between the motions of one side of the component relative to the other. Such relationships may be described by equations like $\{g_i\} = \{g_j\}$ or $1\{g_i\} - 1\{g_j\} = 0$. These equations present the rigid relation between points i and j. Let us gather such equations into a row vector of the form

$$\begin{matrix} 1 & \dots & i & \dots & j & \dots & n \\ \begin{bmatrix} 0 & 0 & 1 & 0 & -1 & 0 & 0 \end{bmatrix} \end{matrix}.$$

The vector has the coefficients of the constraint relationship in the appropriate locations and zeroes elsewhere. For simplicity, in the example above the vector is shown in terms of the node points, but that is not necessary. It is possible to have a rigid relationship between two points only with respect to translations, for example, and not rotations. Then the vector would be extended to six times its size (albeit sparse and not containing the zero terms in commercial implementations) as

$$
\begin{array}{cccccccccccccc}
 & i & & & & & & & & j & & & & \\
\ldots & 1 & 2 & 3 & 4 & 5 & 6 & \ldots & 1 & 2 & 3 & 4 & 5 & 6 \\
\left[\ldots\right. & 1 & 1 & 1 & 0 & 0 & 0 & \ldots & -1 & -1 & -1 & 0 & 0 & 0 & \left.\right].
\end{array}
$$

In practice, many of these constraint conditions may exist; let us assume there is m of them. Then the relationship between the motion of the points of the rigid structure dependent on the adjacent flexible points is described by a rectangular matrix

$$
[R]_{mk} = \left[[R]_{mf} \quad [R]_{mm}\right], \tag{3.67}
$$

where the partition denoted by the subscript f is the collection of points free to move and the remaining m partition's move is dependent on the free points. The correct definition of such a scenario demands that the right submatrix, the collection of dependent degrees of freedom, is invertible. That allows computing the transformation matrix

$$
[G]_{mf} = [R]_{mm}^{-1}[R]_{mf}. \tag{3.68}
$$

We will also partition the input matrices accordingly as

$$
[K]_{kk} = \begin{bmatrix} K_{ff} & K_{fm} \\ K_{mf} & K_{mm} \end{bmatrix}, [D]_{kk} = \begin{bmatrix} D_{ff} & D_{fm} \\ D_{mf} & D_{mm} \end{bmatrix}, [M]_{kk} = \begin{bmatrix} M_{ff} & M_{fm} \\ M_{mf} & M_{mm} \end{bmatrix}.
$$

Here the analysis set of points that are free of any constraints was denoted by the subscript f. The matrix operations

$$
[\bar{K}]_{ff} = [K]_{ff} + [K]_{fm}[G]_{mf} + [G]_{mf}^{T}[K]_{mf} + [G]_{mf}^{T}[K]_{mm}[G]_{mf}, \tag{3.69}
$$

$$
[\bar{D}]_{ff} = [D]_{ff} + [D]_{fm}[G]_{mf} + [G]_{mf}^{T}[D]_{mf} + [G]_{mf}^{T}[D]_{mm}[G]_{mf}, \tag{3.70}
$$

and

$$
[\bar{M}]_{ff} = [M]_{ff} + [M]_{fm}[G]_{mf} + [G]_{mf}^{T}[M]_{mf} + [G]_{mf}^{T}[M]_{mm}[G]_{mf} \tag{3.71}
$$

will produce the free set matrices. The transformation must be carried into the solution displacement vector as

$$\{g\}_f = [G]_{mf}\{g\}_k,\tag{3.72}$$

as well as into the right-hand-side matrix

$$\{F\}_f = [G]_{mf}^T\{F\}_k.\tag{3.73}$$

The remaining set consists of the free degrees of points that are the basis of the analysis equilibrium equation.

We execute a similar partitioning and reduction on the rotor dynamic matrices and produce the $[Z]_{ff},[K_G]_{ff},[K_D]_{ff},[C]_{ff}$ matrices. We introduce the matrices

$$[\tilde{K}]_{ff} = [\bar{K}]_{ff} - \Omega^2[Z]_{ff} + \Omega^2[K_G]_{ff} + \Omega[K_D]_{ff},\tag{3.74}$$

$$[\tilde{D}]_{ff} = [\bar{D}]_{ff} + 2\Omega[C]_{ff},\tag{3.75}$$

$$[M]_{ff} = [\bar{M}]_{ff}.\tag{3.76}$$

With these, the analysis equilibrium equation becomes

$$[M]\{\ddot{g}\} + [\tilde{D}]\{\dot{g}\} + [\tilde{K}]\{g\} = \{F\}.\tag{3.77}$$

We denoted the matrices that are dependent on the rotation speed by the tilde. We omit their subscript notation for simplicity here and in most of the following, except when it is necessary to show.

Equation (3.77) is a unified version of the two (fixed or rotating system) equilibrium equations and will be the basis of the discussion in the following chapters. Equations (3.74) and (3.75) are for the rotating system and need some adjustments for the fixed system that was shown earlier. They both are dependent on the rotation speed as well as possibly time, depending on the coupling scenario. Their content is also dependent on whether the analysis is in a fixed or rotating frame of reference. This, however, is not important for the following discussions, which will be conducted for the generic case.

In practice, there is another set of degrees of freedom that may be constrained by boundary conditions. Their handling is conceptually simple when they are zero boundary conditions; they may simply be removed from the governing equations. When they are nonzero boundary conditions, that is, we enforce some deformations or motions of the system, the process is a bit more involved, but it will not be detailed here. See ref. [1] for more details. In the following two chapters the computational and numerical solutions are discussed in connection with the generic analysis equilibrium equation.

4

Computational Solution Techniques

Most assembled rotor dynamics finite element models result in very large matrices because the structural details are represented with high fidelity. Because the real life operational scenarios also require the solution of the dynamic equations in a wide range of rotational speeds, the accuracy and efficiency of the solution techniques employed are of utmost importance. This chapter mainly focuses on the computational aspects of the solution of the equations developed in the last chapter. The accuracy aspects of the numerical solution of the equations are the topic of the next chapter.

4.1 Direct Time Domain Solution of the Equilibrium Equation

The finite element equilibrium equation is a second-order differential equation in terms of time; hence the direct solution involves time integration. The numerical solution of the equation is obtained at a series of equidistant time steps:

$$t, t + \Delta t, t + 2\Delta t, \ldots.$$

The time integration method most commonly used in industrial finite element analysis is the Newmark method. The method finds the solution in a forward-looking fashion as

$$\{g\}(t + \Delta t) = \beta\{g\}(t) + (1 - 2\beta)\{g\}(t + \Delta t) + \beta\{g\}(t + 2\Delta t). \tag{4.1}$$

(This β is a scalar and has no relation to the $\{\beta\}$ vector introduced in Chapter 2, Equation 2.15.)

This is an implicit method, because the generalized displacement vector at the next time step $t + \Delta t$ is present on both sides. The generalized velocities are computed by the central difference method as

$$\{\dot{g}\}(t + \Delta t) = \frac{\{g\}(t + 2\Delta t) - \{g\}(t)}{2\Delta t}. \tag{4.2}$$

The generalized accelerations by similar means are

$$\{\ddot{g}\}(t + \Delta t) = \frac{\{g\}(t + 2\Delta t) - 2\{g\}(t + \Delta t) + \{g\}(t)}{\Delta t^2}. \tag{4.3}$$

The differential equation at the next time step is

$$[M]\{\ddot{g}\}(t+\Delta t)+[\tilde{D}]\{\dot{g}\}(t+\Delta t)+[\tilde{K}]\{g\}(t+\Delta t)=\{F\}(t+\Delta t). \qquad (4.4)$$

Substituting the generalized velocities and displacements results in the form

$$\frac{[M]}{\Delta t^2}(\{g\}(t+2\Delta t)-2\{g\}(t+\Delta t)+g(t))+\frac{[\tilde{D}]}{2\Delta t}(\{g\}(t+2\Delta t)-\{g\}(t))$$

$$[\tilde{K}](\{g\}(t+2\Delta t)+(1-2\beta)\{g\}(t+\Delta t)+\beta\{g\}(t))=\{F\}(t+\Delta t). \qquad (4.5)$$

The stability of this solution technique is an important issue. This is a distinct issue from the stability of the formulation, which was addressed before. The stability of the Newmark procedure may be derived from the general stability definition of numerical time integration procedures. The stability of any two-step solution scheme for a function f of the form

$$\{g\}(t+2\Delta t)=\{g\}(t+\Delta t)-c_1\{g\}(t+\Delta t)-c_2\{g\}(t) \qquad (4.6)$$

depends on the roots of its characteristic equation, defined by

$$\lambda^2+c_1\lambda+c_2=0. \qquad (4.7)$$

The solution method is stable when the roots of the characteristic equation reside in the complex unit circle, that is, $|\lambda_i|\leq 1, i=1,2$. Analytic evaluation of this criterion for the Newmark method, details of which are beyond our focus but are presented in ref. [1], results in the condition of stability as

$$\beta\geq\frac{1}{4}. \qquad (4.8)$$

The selection of $\beta=1/3$ renders the Newmark solution technique unconditionally stable in the case of constant matrices. Using this value in the reorganization of the above formula provides a time integration process as

$$\left(\frac{[M]}{\Delta t^2}+\frac{[\tilde{D}]}{2\Delta t}+\frac{[\tilde{K}]}{3}\right)\{g\}(t+2\Delta t)=$$

$$\{F\}(t+\Delta t)+\left(\frac{2[M]}{\Delta t^2}-\frac{[\tilde{K}]}{3}\right)\{g\}(t+\Delta t)-\left(\frac{[M]}{\Delta t^2}-\frac{[\tilde{D}]}{2\Delta t}+\frac{[\tilde{K}]}{3}\right)\{g\}(t). \qquad (4.9)$$

The fact that our matrices are a function of the rotation speed makes this issue a bit more delicate. Practical applications of this solution are discussed in Chapter 7, Sections 7.3 and 7.4, dealing with transient solutions.

4.2 Direct Frequency Domain Solution

Another direct solution approach used in the practice is in the frequency domain. Assuming a harmonic solution manifested in the transformation of the form of

$$g(t) = e^{i\omega t} u(\omega), \tag{4.10}$$

and applying it to the time domain equation of

$$[M]\{\ddot{g}\} + [\tilde{D}]\{\dot{g}\} + [\tilde{K}]\{g\}(t) = \{F\}(t), \tag{4.11}$$

results in the frequency domain equilibrium equation

$$(-\omega^2[M] + i\omega[\tilde{D}] + [\tilde{K}])\{u\}(\omega) = \{F\}(\omega). \tag{4.12}$$

Here the load vectors are also harmonic:

$$\{F\}(t) = e^{i\omega t}\{F\}(\omega). \tag{4.13}$$

The advantage of this formulation is in analyses when the load is frequency (i.e., rotation speed) dependent. An example of this is the centrifugal load that we established in earlier chapters as being proportional to Ω^2. There are two major analyses executed in the frequency domain: synchronous and asynchronous.

In the synchronous case we assume that the natural frequency of the rotor and the rotational speed are equal, resulting in the form

$$(-\Omega^2[M] + i\Omega[\tilde{D}] + [\tilde{K}])\{u\}(\Omega) = \{F\}(\Omega). \tag{4.14}$$

The rotational speeds at which such coincidence occurs are called critical speeds. Finding the response at such speeds is of utmost practical importance as the potential resonance scenario may damage the structure. Establishing an operational range outside of the resonance locations is the safest solution. If that cannot be done, the knowledge gained from the synchronous analysis can aid the operational stability by crossing over those frequencies quickly.

The behavior of a rotating structure at the critical speeds depends on the direction of the rotation. If the direction of the load is coincident with the direction of the rotation, the so-called forward whirl motion occurs. Conversely, when the load impacts the structure in a direction opposite to the rotation, the so-called backward whirl occurs.

The synchronous frequency domain analysis is usually executed at certain discrete rotational speeds as

$$\left(-\Omega_k^2[M] + i\Omega_k[\tilde{D}] + [\tilde{K}]\right)\{u\}(\Omega_k) = \{F\}(\Omega_k); k = 1, 2, \dots m. \tag{4.15}$$

The force is dependent on rotor speed. The asynchronous analysis is also a series of solutions, both frequency and rotor speed related. For example, the rotor speed is constant but the force is frequency dependent.

$$\left(-\omega_k^2[M] + i\omega_k[\tilde{D}(\Omega_l)] + [\tilde{K}(\Omega_l)]\right)\{u\}(\omega_k) = \{F\}(\omega_k); k = 1, 2, \dots m; l = 1, 2, \dots n. \tag{4.16}$$

In this equation the rotor speed dependence of the damping and the stiffness matrices, denoted by the tilde, is further quantified in showing the rotor speed variation.

As there are differences in the rotor speed-dependent damping and stiffness matrices whether the fixed or rotational frame of reference is used, there are slight differences in establishing the critical speeds. Such specifics and the interpretation of critical speeds, regions of instability, and whirl motions are the subject of Chapter 6.

4.3 Direct Free Vibration Solution

In the remainder of this chapter the matrices and vectors are not bracketed to avoid cluttering the equations. In general, uppercase characters denote matrices and lowercase characters stand for vectors. Greek characters may be either, but the context will clarify their form. The fundamental problem of rotor dynamics is to find free, unforced vibrations of the rotor when

$$F(\omega) = 0. \tag{4.17}$$

This fundamental engineering problem is mathematically a complex eigenvalue problem:

$$(\lambda^2 M + \lambda \tilde{D} + \tilde{K})\varphi = 0, \tag{4.18}$$

where the λ is a complex eigenvalue and $\varphi = u(\lambda)$ is the complex eigenvector corresponding to λ. The solution of this quadratic eigenvalue problem, of utmost importance in rotor dynamics, is detailed in the following.

The computational solution of this quadratic eigenvalue problem is accomplished by a linearization transformation that results in a single matrix form, called a canonical form. Such transformation may be achieved by using

$$\dot{\varphi} = \lambda \varphi, \tag{4.19}$$

and writing the eigenvalue equation in the following block matrix form:

$$\left(\lambda \begin{bmatrix} M & 0 \\ 0 & I \end{bmatrix} + \begin{bmatrix} \tilde{D} & \tilde{K} \\ -I & 0 \end{bmatrix} \right) \begin{bmatrix} \dot{\varphi} \\ \varphi \end{bmatrix} = 0. \tag{4.20}$$

This form is still not yet amenable for practical analysis; further adjustments are needed. Let us introduce a complex shift of λ_s, a chosen location in the complex plane, and define a computational eigenvalue of

$$\bar{\lambda} = \frac{1}{\lambda - \lambda_s}. \tag{4.21}$$

With this so-called spectral transformation, the problem may be transformed into

$$\left(\begin{bmatrix} M & 0 \\ 0 & I \end{bmatrix} + \bar{\lambda} \begin{bmatrix} \tilde{D} + M\lambda_s & \tilde{K} \\ -I & \lambda_s I \end{bmatrix} \right) \begin{bmatrix} \dot{\varphi} \\ \varphi \end{bmatrix} = 0. \tag{4.22}$$

Finally, a premultiplication by the inverse of the second block matrix yields a single matrix with a corresponding eigenvector as

$$A = \begin{bmatrix} -(\tilde{D} + M\lambda_s) & -\tilde{K} \\ I & -\lambda_s I \end{bmatrix}^{-1} \begin{bmatrix} M & 0 \\ 0 & I \end{bmatrix}, \bar{\varphi} = \begin{bmatrix} \dot{\varphi} \\ \varphi \end{bmatrix}. \tag{4.23}$$

The upper bar over the eigenvectors reflects the fact that the eigenvector underwent a transformation during this process and the physical solutions will have to be recovered.

It is important to note that the matrix of this canonical eigenvalue problem by the nature of the transformation process is unsymmetric; hence, both a conventional and a so-called left-handed solution exists:

$$A \bar{\varphi} = \bar{\lambda} \bar{\varphi}, \bar{\phi}^T A = \bar{\lambda} \bar{\phi}^T. \tag{4.24}$$

When the eigenvectors become complex, the transposes are complex conjugate transposes, but we are not going to distinguish this in notation. The left-handed problem has similar block partitioning to the classical right-handed solution:

$$\begin{bmatrix} \dot{\phi} \\ \phi \end{bmatrix}^T \left(\begin{bmatrix} M & 0 \\ 0 & I \end{bmatrix} + \bar{\lambda} \begin{bmatrix} \tilde{D} + M\lambda_s & \tilde{K} \\ -I & \lambda_s I \end{bmatrix} \right) = 0. \tag{4.25}$$

This fact will be important in recovering the vibration mode shapes of the physical solutions.

The final step in the direct solution is to recover the eigensolution of the original quadratic problem. The eigenvalues are simply computed by reversing the spectral transformation calculation:

$$\lambda = \frac{1}{\bar{\lambda}} + \lambda_s. \tag{4.26}$$

The beauty of this linearization approach is that the right eigenvectors, the characteristic shapes of the displacements, are also invariant. Simple partitioning produces

$$\begin{bmatrix} \dot{\varphi} \\ \varphi \end{bmatrix} = \bar{\varphi}. \tag{4.27}$$

The left eigenvectors, the characteristic shapes of the velocity, are unfortunately not invariant under the transformation. Without going into the details, which are explained in ref. [2], the left physical eigenvectors are computed by

$$\begin{bmatrix} \dot{\phi} \\ \phi \end{bmatrix} = - \begin{bmatrix} -(\tilde{D} + \lambda_s M) & -\tilde{K} \\ I & -\lambda_s I \end{bmatrix}^{-1,T} \bar{\phi}. \tag{4.28}$$

The importance of computing the left physical vectors is in the fact that they provide a means to evaluate the computational solution. Let us consider the *i*-th right-hand solution of the rotor dynamics eigenvalue problem,

$$(\lambda_i^2 M + \lambda_i \tilde{D} + \tilde{K})\varphi_i = 0, \tag{4.29}$$

and the *j*-th left-hand solution of

$$\phi_j^T (\lambda_j^2 M + \lambda_j \tilde{D} + \tilde{K}) = 0. \tag{4.30}$$

With appropriate pre- and postmultiplication and some linear algebraic tedium, one can obtain two orthogonality conditions:

$$\phi_j^T M \phi_i (\lambda_i + \lambda_j) + \phi_j^T \tilde{D} \phi_i = D_1 \tag{4.31}$$

and

$$\lambda_j \phi_j^T M \phi_i \, \lambda_i - \phi_j^T \tilde{K} \phi_i = D_2. \tag{4.32}$$

The matrices on the right-hand sides are diagonal apart from computational zeroes in the off-diagonal terms. The diagonal terms of the D_1 matrix are λ_i, and the diagonal of the D_2 matrix are of the order of the λ_i^2 quantities. The first matrix measures the orthogonality relationship between the mass and the damping matrices, and the second matrix expresses the quality of the mass and stiffness matrix relationship.

The eigenvalue solution has a significant role in computing the reduced-order modal solutions of the finite element equilibrium equation, which is the subject of the next sections. Besides that, the complex eigenvalues of the rotating structure provide the most important engineering analysis components: the evaluation of critical speeds and the regions of stability. These will be discussed in Chapter 6.

4.4 Modal Solution Technique

To address the modal solution of a rotor dynamic response problem, we consider the time domain forced response problem of the free set degrees of freedom:

$$M_{ff} \ddot{u}_f + \tilde{D}_{ff} \dot{u}_f + \tilde{K}_{ff} u_f = F_f. \tag{4.33}$$

Note that we returned to the subscripts of the matrices as they are instrumental in explaining this solution technique. The modal solution is based on the real, symmetric eigenvalue problem of

$$\left(M_{ff} \lambda_i^2 + K_{ff} \right) \varphi_{fi} = 0; i = 1, \dots h. \tag{4.34}$$

Note that the tilde over the stiffness matrix is not present in this equation, demonstrating the fact that the rotational speed-dependent matrices augmenting the stiffness matrix are not considered. In other words, we solve the simple unrotating structural problem. Let the eigenvectors be concatenated into the array

$$\Phi_{fh} = \begin{bmatrix} \varphi_{f1} & \cdots & \varphi_{fh} \end{bmatrix}, \tag{4.35}$$

where h is the number of eigenvectors captured. We introduce the modal solution as a transformation by the eigenvector matrix:

$$u_f = \Phi_{fh}u_h. \tag{4.36}$$

Substitution into Equation (4.33) and multiplication from the left results in

$$\Phi_{fh}^T(M_{ff}\Phi_{fh}\ddot{u}_h + \tilde{D}_{ff}\Phi_{fh}\dot{u}_h + \tilde{K}_{ff}\Phi_{fh}u_h) = \Phi_{fh}^T F_f. \tag{4.37}$$

The modal solution equation is of the form

$$m_{hh}\ddot{u}_h + \tilde{d}_{hh}\dot{u}_h + \tilde{k}_{hh}u_h = f_h. \tag{4.38}$$

Note that the modal matrices, apart from the modal mass, are not diagonal because the rotational effects are also represented, denoted by the tilde over the modal stiffness and the modal damping. The modal matrices are computed as

$$m_{hh} = \Phi_{fh}^T M_{ff}\Phi_{fh}, \tilde{b}_{hh} = \Phi_{fh}^T \tilde{D}_{ff}\Phi_{fh}, \tilde{k}_{hh} = \Phi_{fh}^T \tilde{K}_{ff}\Phi_{fh}, f_h = \Phi_{fh}^T F_f. \tag{4.39}$$

The major shortcoming of the approach is that the limited number of eigenvectors (h) representing the modal space of the problem results in the so-called modal space truncation. Let us view the complete modal space, that is, the set of all eigenvectors, and partition as

$$\Phi_{ff} = \begin{bmatrix} \Phi_{fh} & \Phi_{fr} \end{bmatrix}. \tag{4.40}$$

Here $r = f - h$ is the size of the second partition containing the remaining eigenvectors that are not computed. To measure the effect of the truncation, we consider the inverse of the stiffness matrix. Due to normalization to unit modal mass, the following relationship holds:

$$K_{ff}^{-1} = \Phi_{ff}\Lambda_{ff}^{-1}\Phi_{ff}^T. \tag{4.41}$$

Here the middle matrix holds all the eigenvalues on its diagonal. Using the previous partitioning, this may be written as

$$K_{ff}^{-1} = \Phi_{fh} \Lambda_{hh}^{-1} \Phi_{fh}^T + \Phi_{fr} \Lambda_{rr}^{-1} \Phi_{fr}^T = Z_h + Z_r, \tag{4.42}$$

where the terms on the right-hand side are the flexibilities carried in the computed modal space and in the remaining modal space, respectively. The flexibility of the remaining uncomputed modal space, commonly called the residual flexibility, is then

$$Z_r = K_{ff}^{-1} - \Phi_{fh} \Lambda_{hh}^{-1} \Phi_{fh}^T. \tag{4.43}$$

A possible way to account for the missing residual flexibility is by introducing static solution vectors responding to unit loads placed on certain points of the structure. Assuming $k \le r$ for such loads, the static vectors are computed as

$$\tilde{\Psi}_{fk} = Z_r U_k, \tag{4.44}$$

where the U matrix holds the columns of the static loads. Substituting yields the formula

$$\tilde{\Psi}_{fk} = K_{ff}^{-1} U_k - \Phi_{fh} \Lambda_{hh}^{-1} \Phi_{ff}^T U_k, \tag{4.45}$$

where all the terms on the right-hand side are known, or computable. The inverse of the stiffness matrix is, of course, never computed in practical circumstances, and the solution of a linear system produces the first term. After orthogonalizing the newly computed vectors against the existing eigenvectors, the so-obtained Ψ_{fk} vectors will augment the computed modal space as

$$\Phi_{fl} = \begin{bmatrix} \Phi_{fh} & \Psi_{fk} \end{bmatrix}, \tag{4.46}$$

where $l = h + k$. The modal matrices may be recomputed as

$$m_{ll} = \Phi_{fl}^T M_{ff} \Phi_{fl}, \tilde{b}_{ll} = \Phi_{fl}^T \tilde{D}_{ff} \Phi_{fl}, \tilde{k}_{ll} = \Phi_{fl}^T \tilde{K}_{ff} \Phi_{fl}, f_k = \Phi_{fl}^T F_f. \tag{4.47}$$

The location of the unit load points is not random and needs to be judiciously chosen by an engineer knowledgeable about the structure. Hence this augmentation of the modal space, while provided as an option in advanced commercial software, is an expert user topic.

The previous modal formulation is focused on the time domain, but the frequency domain formulation also follows simply in the form of

$$\left(-\omega^2 m_{hh} + i\omega\, \tilde{d}_{hh} + \tilde{k}_{hh}\right) u_h(\omega) = f_h(\omega). \tag{4.48}$$

The modal space augmentation process outlined above may also be applied to this case. In either case, the reduced-order equation does not carry the full dynamic content of the problem as was posed above. The accuracy of the solution depends on the details of the structural model. Some simpler, axially symmetric rotor components like shafts may be accurately analyzed by the real modal solution technique. Models with more elaborate structural details, such as turbine wheels with a multitude of blades, require better methods; these are addressed below.

When the internal damping of the structure cannot be ignored, represented by the D_I matrix in Chapter 3, the real modal solution is inaccurate. Furthermore, when the effect of the bearing cannot be ignored, there are two additional matrices to consider. There is the external damping matrix, denoted by D_E, and the unsymmetric stiffness matrix component K_D, whose participation requires a complex modal solution. It is based on the quadratic eigenvalue problem of

$$\left(M_{ff}\lambda^2 + D_{ff}\lambda + \hat{K}_{ff}\right)\varphi_f = 0. \tag{4.49}$$

Here the matrix $D_{ff} = D_I + D_E$ and $\hat{K}_{ff} = \bar{K}_{ff} + \Omega^* K_D$. The notation of the damping matrix implies that the rotation speed-dependent Coriolis damping component is not considered. The notation of the stiffness matrix hints at some limited dependence on the rotation speed. Because the structural damping term is multiplied by the rotation speed in the equilibrium equation, but the modal reduction is executed *a priori* to the RPM loop, we compute a static value of this structural stiffness at a certain rotation speed Ω^*. That speed could be the lowest rotational speed or a medial value in the speed interval.

Such a problem, of course, has a left-handed solution as well:

$$\phi_f^T\left(M_{ff}\lambda^2 + D_{ff}\lambda + \hat{K}_{ff}\right) = 0. \tag{4.50}$$

Let the left and right eigenvectors be concatenated into the arrays

$$\Psi_{fh} = \begin{bmatrix} \phi_{f1} & \cdots & \phi_{fh} \end{bmatrix} \tag{4.51}$$

and

$$\Phi_{fh} = \begin{bmatrix} \varphi_{f1} & \cdots & \varphi_{fh} \end{bmatrix}, \tag{4.52}$$

respectively, where h is the number of (possibly complex) eigenvector pairs. We introduce the complex modal solution as

$$u_f = \Phi_{fh} u_h. \tag{4.53}$$

Substitution into Equation (4.33) and subsequent multiplication by the left eigenvector results in

$$\Psi_{fh}^T (M_{ff} \Phi_{fh} \ddot{u}_h + \tilde{B}_{ff} \Phi_{fh} \dot{u}_h + \tilde{K}_{ff} \Phi_{fh} u_h) = \Psi_{fh}^T F_f. \tag{4.54}$$

The complex modal solution equation is of the form

$$m_{hh} \ddot{u}_h + \tilde{b}_{hh} \dot{u}_h + \tilde{k}_{hh} u_h = f_h, \tag{4.55}$$

where

$$\Psi_{fh}^T M_{ff} \Phi_{fh} = m_{hh}, \, \Psi_{fh}^T \tilde{B}_{ff} \Phi_{fh} = \tilde{b}_{hh}, \, \Psi_{fh}^T \tilde{K}_{ff} \Phi_{fh} = \tilde{k}_{hh}, \, p_h = \Psi_{fh}^T F_f. \tag{4.56}$$

Note, however, that none of these modal matrices are diagonal anymore, but the complex modal form carries a higher fidelity of the original problem as opposed to the real modal form.

4.5 Static Condensation

The static condensation utilizes a partitioning of the matrices into two partitions, one denoted by the subscript o (omitted) and another by the subscript a (analyzed). Such a partitioning of the stiffness matrix is

$$K_{ff} = \begin{bmatrix} K_{oo} & K_{oa} \\ K_{ao} & K_{aa} \end{bmatrix}. \tag{4.57}$$

The mass and damping matrices undergo a similar partitioning. Note again that the stiffness matrix under consideration does not contain the rotational effects—it is symmetric. We introduce a condensation matrix

$$G_{oa} = -K_{oo}^{-1} K_{oa}. \tag{4.58}$$

The process called static condensation of the global stiffness matrix is executed as

$$
\begin{bmatrix} I_{oo} & 0 \\ G_{ao} & I_{aa} \end{bmatrix} \begin{bmatrix} K_{oo} & K_{oa} \\ K_{ao} & K_{aa} \end{bmatrix} \begin{bmatrix} I_{oo} & G_{oa} \\ 0 & I_{aa} \end{bmatrix} = \begin{bmatrix} K_{oo} & 0 \\ 0 & \bar{K}_{aa} \end{bmatrix}.
\tag{4.59}
$$

Here the modified a-partition stiffness matrix is of the form

$$
\bar{K}_{aa} = K_{ao}G_{oa} + K_{aa} = K_{aa} - K_{ao}K_{oo}^{-1}K_{oa}.
\tag{4.60}
$$

This is the well-known mathematical Schur complement, and the interior partition (K_{oo}) of the stiffness matrix is unchanged. Similar transformations on the mass matrix will not result in a block diagonal form, but in the form of

$$
\begin{bmatrix} M_{oo} & \bar{M}_{oa} \\ \bar{M}_{ao} & \bar{M}_{aa} \end{bmatrix},
\tag{4.61}
$$

retaining its coupling components in the reduced form of

$$
\bar{M}_{oa} = M_{oo}G_{oa} + M_{aa},
\tag{4.62}
$$

and with $G_{ao} = G_{oa}^T$

$$
\bar{M}_{ao} = G_{ao}M_{oo} + M_{ao}.
\tag{4.63}
$$

The a-partition of the mass matrix is not as simple as that of the stiffness matrix:

$$
\bar{M}_{aa} = G_{ao}M_{oo}G_{oa} + G_{ao}M_{oa} + M_{ao}G_{oa} + M_{aa},
\tag{4.64}
$$

because the static condensation effect does not apply. A similar transformation may be executed on the damping matrix that is not detailed here. The statically condensed damping matrix is identical in form to that of the mass matrix except for the replacement of the M terms with the corresponding B partitions.

The physical loads are also condensed as

$$
\begin{bmatrix} F_o \\ \bar{F}_a \end{bmatrix} = \begin{bmatrix} I_{oo} & 0 \\ G_{ao} & I_{aa} \end{bmatrix} \begin{bmatrix} F_o \\ F_a \end{bmatrix},
\tag{4.65}
$$

where $\bar{F}_a = G_{ao}F_o + F_a$.

Introducing the previous condensed matrices into Equation (4.33) does not result in a reduced-order form, as in the modal solution. This only produces a reduced density form, due to the block diagonalization of the stiffness matrix. This still may provide some solution performance advantage in certain cases in practice, because the stiffness matrix is usually the densest of the three matrices.

The relation between the original physical solution and the statically condensed solution is based on the transformation

$$u_f = \begin{bmatrix} u_o \\ u_a \end{bmatrix} = \begin{bmatrix} I_{oo} & G_{oa} \\ 0 & I_{aa} \end{bmatrix} \begin{bmatrix} \bar{u}_o \\ u_a \end{bmatrix}. \tag{4.66}$$

The equation demonstrates that the statically condensed solution's interior component is back-transformed as $u_0 = \bar{u}_o + G_{oa}u_a$.

The static condensation method has significant disadvantages in dynamic solutions because the interior partitions' masses are only statically represented. To overcome this disadvantage the dynamic reduction method is applied.

4.6 Dynamic Reduction

This method is so named because it dynamically reduces the interior partition. First, we still consider the case when the stiffness matrix is symmetric. This is accomplished by the execution of a real eigenvalue solution on the interior partition as

$$(M_{oo}\lambda_i^2 + K_{oo})\varphi_{oi} = 0; i = 1,...m. \tag{4.67}$$

We use the fixed boundary assumption. For example, for this eigenvalue solution the boundary displacements are assumed to be zero. Note again the lack of tilde over the stiffness matrix indicating the nonrotational components. Let us concatenate the eigenvectors and eigenvalues into two arrays

$$\Lambda_{mm} = \begin{bmatrix} \lambda_1 & & \\ & \cdot & \\ & & \lambda_m \end{bmatrix}, \Phi_{om} = \begin{bmatrix} \varphi_{o1} & \cdots & \varphi_{om} \end{bmatrix}. \tag{4.68}$$

Combining this with the static condensation transformation of the last section we obtain a dynamic reduction transformation matrix as

$$T_{fh} = \begin{bmatrix} \Phi_{om} & G_{oa} \\ 0 & I_{aa} \end{bmatrix}. \tag{4.69}$$

Exploiting the common mass-orthonormality convention of eigenvalue solutions, the dynamically reduced matrices will become

$$T_{fh}^T \tilde{K}_{ff} T_{fh} = \tilde{K}_{hh},$$

$$T_{fh}^T M_{ff} T_{fh} = M_{hh}. \tag{4.70}$$

Here the h size now is the sum of the a-partition plus the number of modes chosen to represent the dynamic behavior of the interior. The interior components of the reduced matrices are now not condensed as in the last section; however, the overall problem size has been significantly reduced. The reduction of the rotation-dependent damping matrix is

$$T_{fh}^T \tilde{D}_{ff} T_{fh} = \tilde{D}_{hh}. \tag{4.71}$$

The dynamically reduced solution is of a different size because

$$u_f = \begin{bmatrix} u_o \\ u_a \end{bmatrix} = \begin{bmatrix} \Phi_{om} & G_{oa} \\ 0 & I_{aa} \end{bmatrix} \begin{bmatrix} u_m \\ u_a \end{bmatrix} = T_{fh} u_h. \tag{4.72}$$

Similarly, the reduced loads are

$$F_h = \begin{bmatrix} \Phi_{om}^T & 0 \\ G_{oa}^T & I_{aa} \end{bmatrix} F_f. \tag{4.73}$$

Introducing the above into Equation (4.33) results in the dynamically reduced-order response problem of

$$M_{hh} \ddot{u}_h + \tilde{D}_{hh} \dot{u}_h + \tilde{K}_{hh} u_h = F_h. \tag{4.74}$$

After solving this equation the physical solution is recovered by Equation (4.72). Depending on the number of modes (m) chosen to reduce the interior, the size reduction could be immense. We will review a case study of this in Chapter 8.

A complex variation of this method is also possible with a complex interior eigenvalue solution, again for the case when the damping cannot be ignored and the stiffness matrix is unsymmetric. In this case the off-diagonal blocks are not transposes of each other:

$$K_{ao} \neq K_{oa}^T.$$

This results in two distinct condensation matrices:

$$G_{oa} = -K_{oo}^{-1}K_{oa}, \; G_{ao} = -K_{ao}K_{oo}^{-1}, \tag{4.75}$$

and they are not each other's transposes, that is, $G_{ao} \neq G_{oa}^T$.

The complex eigensolution will produce different left and right eigenvectors, and we again use the fixed boundary assumption. The right-hand transformation matrix is the same as in the real dynamic reduction and is of the form

$$T_{fh} = \begin{bmatrix} \Phi_{om} & G_{oa} \\ 0 & I_{aa} \end{bmatrix}. \tag{4.76}$$

The left-hand transformation matrix is of the form

$$R_{hf} = \begin{bmatrix} \Psi_{om}^T & 0 \\ G_{ao} & I_{aa} \end{bmatrix}. \tag{4.77}$$

In the above equations the h size again is the sum of the a-partition size and the number of eigenvector pairs forming the modal space, m. The transformation of the stiffness matrix is now executed as follows:

$$R_{hf} \tilde{K}_{ff} T_{fh} = \tilde{K}_{hh}. \tag{4.78}$$

Again, the transformed stiffness matrix is not block diagonal as in static condensation, but it is still significantly reduced in order. Similar transformation is executed on the damping and mass matrices:

$$R_{hf} \tilde{D}_{ff} T_{fh} = \tilde{D}_{hh}, \; R_{hf} M_{ff} T_{fh} = M_{hh}. \tag{4.79}$$

The relation between the physical and the reduced-order solution is based on the right-hand transformation matrix as

$$u_f = \begin{bmatrix} \Phi_{om} & G_{oa} \\ 0 & I_{aa} \end{bmatrix} \begin{bmatrix} u_m \\ u_a \end{bmatrix} = T_{fh} u_h. \tag{4.80}$$

The physical loads are transformed by the left transformation:

$$F_h = \begin{bmatrix} \Psi_{om}^T & 0 \\ G_{ao} & I_{aa} \end{bmatrix} \begin{bmatrix} F_o \\ F_a \end{bmatrix} = R_{hf} F_f. \tag{4.81}$$

Introducing the previous equations into Equation (4.33) results in the reduced-order response problem of

$$M_{hh}\ddot{u}_h + \tilde{D}_{hh}\dot{u}_h + \tilde{K}_h u_h = F_h. \tag{4.82}$$

That is formally the same as Equation (4.74), but the content of the matrices is different and the fidelity of the solution is much higher. After solving this equation, the physical solution is recovered by Equation (4.80).

5

Numerical Solution Techniques

The numerically accurate solution of the fundamental complex eigenvalue problem is discussed in the first section of this chapter. This will be followed by two important components of the method, and the chapter concludes with the solution of periodic equations.

5.1 The Lanczos Method

The accurate numerical solution of large eigenvalue problems is best executed by the well-known Lanczos method described in ref. [2]. The method, originally published more than 60 years ago, became the *de facto* industry standard solution of such problems.

The method is based on the original idea of Cornelius Lanczos to find the characteristic polynomial of an unsymmetric matrix by generating a sequence of vectors resulting in a successive set of polynomials. The process starts from a pair of random vectors (p_0, q_0), and a pair of new vectors is chosen by a linear combination of the starting pair as

$$p_1 = A^T p_0 - \alpha_0 p_0; \; q_1 = A q_0 - \alpha_0 q_0, \tag{5.1}$$

where the A matrix is the matrix of Chapter 4, Equation (4.23).

The coefficient of the linear combination is selected to minimize the magnitude of the new vectors. This may be formally written as

$$p_1^2 = (A^T p_0 - \alpha_0 p_0)^2 = \min; \; q_1^2 = (A q_0 - \alpha_0 q_0)^2 = \min. \tag{5.2}$$

Differentiation of these expressions leads to the simultaneous achievement of these minima by selecting

$$\alpha_0 = \frac{p_0^T (A q_0)}{p_0^T q_0} = \frac{(A^T p_0) q_0}{p_0^T q_0} = p_0^T A q_0. \tag{5.3}$$

This simple common form reflects the fact that the starting vectors are normalized to unit length, that is, $|p_0|,|q_0|=1$. Then the new pair of vectors is also orthogonal to the prior pair, which can be expressed in the equations

$$p_1^T p_o = q_1^T q_0 = 0. \tag{5.4}$$

The process may be continued for another pair defined by the same principle as

$$p_2 = A^T p_1 - \alpha_1 p_1 - \gamma_0 p_0,$$

$$q_2 = A q_1 - \alpha_1 q_1 - \beta_0 q_0. \tag{5.5}$$

The earlier introduced coefficient of the linear combination is selected similarly, but now in terms of the second pair of vectors:

$$\alpha_1 = \frac{p_1^T(Aq_1)}{p_1^T q_1} = \frac{(A^T p_1)q_1}{p_1^T q_1} = p_1^T A q_1. \tag{5.6}$$

Here the formula again reflects the fact that the vectors are normalized to unit length. The two newly introduced coefficients are defined as the cross-combinations of

$$\beta_0 = \frac{p_0^T(Aq_1)}{p_0^T q_1}; \ \gamma_0 = \frac{(A^T p_1)q_0}{p_1^T q_0}. \tag{5.7}$$

Lanczos's brilliant recognition was that due to the ongoing orthogonality relations, the process may be continued, and at any stage there are only three active pairs of vectors in the process. Hence this became known as Lanczos's three-member recurrence. Further computational simplification may be achieved by not transposing the matrix, but by employing a left multiplication. With these, the generic stage of the Lanczos recurrence is of the form

$$\bar{q}_{k+1} = A q_k - \alpha_k q_k - \gamma_{k-1} q_{k-1},$$

$$\bar{p}_{k+1}^T = p_k^T A - \alpha_k p_k^T - \beta_{k-1} p_{k-1}^T. \tag{5.8}$$

This pair of equations may be recursively applied until the order of the matrix is exhausted and infinite precision of arithmetic is assumed. The computation of the k-th incarnation of the first coefficient is a simple generalization of the above as

$$\alpha_k = p_k^T A q_k. \tag{5.9}$$

Unfortunately, the second two coefficients are not uniquely defined. Any combination of them is satisfactory as long as the relation holds:

$$\beta_k \gamma_k = \bar{p}_{k+1}^T \bar{q}_{k+1}. \tag{5.10}$$

A simple selection is such as to achieve the unit normalization of the new vectors desired by the recurrence

$$\beta_k = |\bar{p}_{k+1}|; \ \gamma_k = |\bar{q}_{k+1}|. \tag{5.11}$$

This selection indicates why, in the finite precision environment of computers, the process may break down before reaching the order of the matrix. It may be restarted in practice, but these implementation issues are beyond our focus.

The beauty of the Lanczos method is that the collection of the orthogonal vector pairs, when gathered into the matrices

$$P = \begin{bmatrix} p_0 & p_1 & \cdot & p_{n-1} \end{bmatrix}; \ Q = \begin{bmatrix} q_0 & q_1 & \cdot & q_{n-1} \end{bmatrix}, \tag{5.12}$$

will formally produce a tridiagonal reduction of the original matrix:

$$P^T A Q = T. \tag{5.13}$$

Intriguingly, the P, Q matrices do not have to be explicitly formed and the multiplication need not be explicitly executed, because the content of the tridiagonal matrix is simply the coefficients of the recurrence process

$$T = \begin{bmatrix} \alpha_0 & \gamma_0 & & & \\ \beta_0 & \alpha_1 & \gamma_1 & & \\ & \cdot & \cdot & \cdot & \\ & & \beta_{n-3} & \alpha_{n-2} & \gamma_{n-2} \\ & & & \beta_{n-2} & \alpha_{n-1} \end{bmatrix}. \tag{5.14}$$

Finding the eigenvalues of the tridiagonal form is now simple and may even be accomplished by the classical means of evaluating the determinant

$$\det(T - \lambda I) = 0. \tag{5.15}$$

Because the selection of the off-diagonal terms has some freedom, they may even be selected so that the subdiagonal consists of unit values, further simplifying the evaluation of the determinant. Furthermore, the β_k, γ_k

coefficients may also be chosen so that, despite the fact that the original matrix was unsymmetric, the tridiagonal matrix becomes symmetric, apart from the sign of the off-diagonal coefficients.

In industrial practice, however, the eigenvalues of the tridiagonal matrix are obtained by the procedure known as QR iteration. In this procedure the shifted tridiagonal matrix is factorized as follows:

$$T - \kappa I = O^0 R^0, \tag{5.16}$$

where the O^0 matrix contains orthonormal columns and the R^0 matrix is upper triangular. Note that the O^i matrices of this procedure are usually denoted by Q to adhere to the notation captured in the name of the procedure. In order to distinguish from the Q matrix of the Lanczos procedure, we will use the O notation for the orthogonal matrix of the above factorization. The details of such an orthogonal factorization are discussed in the next section.

After the execution of the factorization, the iteration proceeds by computing a new version of the tridiagonal matrix

$$T^1 = R^0 O^0 + \kappa I. \tag{5.17}$$

Because the columns of the O^0 matrix are orthonormal, it follows that the

$$T^1 = O^{0,T} T O^0 \tag{5.18}$$

matrix preserves the eigenvalues of the earlier one. The above two steps are repeated as

$$T^k - \kappa I = O^k R^k; \; T^{k+1} = R^k O^k + \kappa I, \tag{5.19}$$

until the next iterate of the matrix becomes computationally diagonal (numerically negligible off-diagonal terms). At that stage the diagonal terms are the eigenvalues of the matrix

$$T^k \cong diag(\lambda). \tag{5.20}$$

The solution for the eigenvectors of the matrix is accomplished in two phases. First, the eigenvectors of the tridiagonal matrix, with now known eigenvalues, may be found from the pair of homogeneous equations

$$(T - \lambda I)v; \; u^T (T - \lambda I) = 0. \tag{5.21}$$

The solution may exploit the fact that the eigenvectors are unique only to a scalar multiplier. The expressions demonstrate the fact that the eigenvalues are invariant under the tridiagonal transformation. The eigenvectors of the original A matrix are recovered by

$$\bar{\phi} = P^T u; \quad \bar{\phi} = Qv, \tag{5.22}$$

yielding the final solution of the original eigenvalue problem as

$$A\bar{\phi} = \lambda\bar{\phi}; \quad \bar{\phi}^T A = \bar{\phi}^T \lambda. \tag{5.23}$$

There are many other numerical considerations in finding the eigenvalues of the reduced form, such as the potential presence of multiple eigenvalues. This is commonplace in engineering applications due to the symmetries inherent in structures, and this is especially so in rotor dynamics.

5.2 Orthogonal Factorization

The orthogonal factorization of a matrix is the central operation of the iteration of the prior section. Specifically, this factorization may be accomplished with a systematic orthogonalization of the columns of the matrix on the left-hand side.

Note that the orthogonal factorization is applicable to rectangular matrices also—a fact of significance in the following. Let us assume that a matrix to be factorized in such a fashion has n rows and m columns, where the number of columns is equal to or less than the number of rows. Let us factorize a matrix C, whose columns are designated by the column vectors $c_k; k = 1, 2, ..., m$ and the columns of O are designated by the vectors $o_k; k = 1, 2, ..., m$. Then the orthogonal factorization of the matrix column-wise is represented by

$$c_k = \sum_{i=1}^{k} r_{ik} o_i; k = 1, 2, \ldots, m, \tag{5.24}$$

where $r_{ik}; i = 1, 2, ..., m$ are the scalar terms of the R matrix whose dimensions are m rows and m columns. Assuming that the C matrix has full column rank (a valid assumption for our application), all the diagonal r_{kk} terms are nonzero and the following equation may be written

$$o_k = \frac{1}{r_{kk}} \left(c_k - \sum_{i=1}^{k-1} \left(o_i^T c_k \right) o_i \right). \tag{5.25}$$

This equation is well known as the Gram–Schmidt orthogonalization scheme. The term in the parenthesis of the summation is r_{ik}; hence the formula may be rewritten as

$$o_k = \frac{1}{r_{kk}} \left(c_k - \sum_{i=1}^{k-1} r_{i,k} o_i \right). \tag{5.26}$$

The computational process starts from $k = 1$ computing $r_{11} = \sqrt{c_1^T c_1}$ and

$$o_1 = \frac{1}{r_{11}} c_1.$$

The process continues with $k = 2$, $r_{12} = \sqrt{o_1^T c_2}$ and computing an intermediate quantity

$$v_2 = c_2 - \sum_{i=1}^{1} r_{12} o_1.$$

This vector produces the next diagonal term, $r_{22} = \sqrt{v_2^T v_2}$, and the next column of the O matrix as

$$o_2 = \frac{1}{r_{22}} v_2.$$

The process is easily generalized for all values of k until m, resulting in the desired factorization

$$C = \begin{bmatrix} c_1 & c_2 & \cdot & c_m \end{bmatrix} = \begin{bmatrix} o_1 & o_2 & \cdot & o_m \end{bmatrix} \begin{bmatrix} r_{11} & r_{12} & \cdot & r_{1m} \\ 0 & r_{22} & \cdot & r_{2m} \\ 0 & 0 & \cdot & \cdot \\ 0 & 0 & 0 & r_{mm} \end{bmatrix} = OR. $$

$$\tag{5.27}$$

The orthogonal factorization is also a cornerstone of the block Lanczos method, detailed in the next section.

5.3 The Block Lanczos Method

The robust numerical solution of our quadratic eigenvalue problem is obtained by a specific advanced version of the block Lanczos method. The advantage of the method is in its ability to capture the multiple eigenvalues commonly occurring in practical structures due to symmetry conditions. The method is based on the generalization of Equation (5.8) as

$$\bar{P}_{j+1}^T = P_j^T A - A_j P_j^T - C_j P_{j-1}^T \tag{5.28}$$

and

$$\bar{Q}_{j+1} = A Q_j - Q_j A_j - Q_{j-1} B_j. \tag{5.29}$$

A certain number of the left-hand vectors denoted by p_j in the last section are collected into the block P_j, and similarly the right-hand vectors q_j of the prior process are gathered into Q_j. The A_j submatrix takes the role of the α_j coefficients, and the B_j lower triangular submatrix will represent the β_j coefficients. Finally, the upper triangular C_j blocks replace the γ_j coefficients.

To obtain the next set of left and right Lanczos blocks, the orthogonal factorization described in the last section is employed as

$$\bar{P}_{j+1}^T = B_{j+1} P_{j+1}^T; \; \bar{Q}_{j+1} = Q_{j+1} C_{j+1}, \tag{5.30}$$

where the Q_{j+1} matrix is the O orthogonal and C_{j+1} is the R triangular matrix of the second factorization. Similar relations hold for the first factorization with consideration of the transpose, that is, the P_{j+1} matrix is the O matrix and the B_{j+1}^T matrix is the triangular R matrix.

The recursive process described in Section 5.1 with the single vector method is similarly executed with the above block operations. Note that the original matrix is only used as a multiplier in both block equations; it is neither transposed nor updated. As pointed out before and is highly visible in this formulation, this is one of the most powerful aspects of the method.

In practical problems consisting of millions of degrees of freedom, the eigenvalue extraction is not executed until all the eigenvalues are found. It is impractical in computational sense and unnecessary for the engineering problem. In practical circumstances the eigenvalue reduction is executed only up to a certain number of eigenvalues dictated by the application area.

Assuming that j block steps were executed, the left and right Lanczos vector blocks are gathered up into the arrays

$$\bar{P} = \begin{bmatrix} P_1 & P_2 & . & P_j \end{bmatrix}$$

and

$$\bar{Q} = \begin{bmatrix} Q_1 & Q_2 & . & Q_j \end{bmatrix}.$$

These blocks of matrices would formally reduce the original matrix to a block tridiagonal form by the following multiplication operations:

$$T_b = \bar{P}^T A \bar{Q}. \tag{5.31}$$

Fortunately, again, this multiplication is not executed explicitly, because the structure of the resulting block tridiagonal matrix has the following content:

$$T_b = \begin{bmatrix} A_1 & B_1 & & & \\ C_1 & A_2 & B_2 & & \\ & C_2 & \cdot & \cdot & \\ & & \cdot & \cdot & B_j \\ & & & C_j & A_j \end{bmatrix}. \tag{5.32}$$

The number of actual rows in the matrix is the number of block Lanczos steps executed multiplied by the block size, which is usually constant. In some cases an adjustment of the block size is necessitated by numerical considerations, but the details of such are beyond our focus. Leading commercial software implementations incorporate such capability.

Just as was mentioned above, the eigenvalues are invariant under the block tridiagonal transformation as well, and the eigenvalue solution of the block tridiagonal matrix will produce computational approximations of the eigenvalues of the original matrix. Then the left- and right-hand eigenvectors of the tridiagonal problem

$$u^T T_b = \lambda u^T , \, T_b v = \lambda v \tag{5.33}$$

are transformed to become the eigenvectors of the original matrix as

$$\bar{\phi} = \bar{P}^T u, \bar{\varphi} = \bar{Q} v. \tag{5.34}$$

The bar over the eigenvectors indicates that these are those of the canonical problem and not of the original quadratic problem. Chapter 4, Section 4.3 discussed the recovery of the eigenvectors of the original rotor dynamics problem.

These steps conclude the numerical solution of the canonical eigenvalue problem.

5.4 Solution of Periodic Equations

Thus far we have only focused on the solution of the nonperiodic finite element equilibrium equation. The more difficult coupling scenarios introduced in Chapter 2, Sections 2.3 and 2.4 result in equilibrium equations of periodic nature. The solution of periodic equations requires some additional considerations. The coupled equilibrium equation

$$[M(t)]\{\ddot{g}\} + [\tilde{D}(t)]\{\dot{g}\} + [\tilde{K}(t)]\{g\} = \{F(t)\} \tag{5.35}$$

now has terms in the matrices that contain trigonometric functions of time and is hence periodic. Note that for clarity we reverted to the bracket notations that we abandoned at one point in the beginning of the last chapter for simplicity.

For the first part of this discussion we ignore the right-hand-side load and transform our second-order periodic equation into a first-order equation in the state variable

$$\{\dot{h}(t)\} = [A(t)]\{h(t)\}. \tag{5.36}$$

The components of this matrix equation are

$$[A(t)] = \begin{bmatrix} -[M(t)]^{-1}[\tilde{D}(t)] & -[M(t)]^{-1}[\tilde{K}(t)] \\ [I] & [0] \end{bmatrix}; \{h(t)\} = \begin{Bmatrix} \{\dot{g}(t)\} \\ \{g(t)\} \end{Bmatrix}. \tag{5.37}$$

From the periodic nature of the system it is true that

$$[A(t)] = [A(t+T)]. \tag{5.38}$$

The solution of such a periodic system may be found in the form proposed by the Frenchman Floquet about hundred years ago,

$$\{h(t)\} = [\Psi(t)][e^{\eta\,t}], \tag{5.39}$$

where $[\Psi(t)]$ are periodic eigenvectors with a period of T; hence $[\Psi(0)] = [\Psi(T)]$. The matrix of $[e^{\eta t}]$ exponential terms is a diagonal matrix. Both of these are related to the eigenvalues and eigenvectors of a matrix $[\Phi(t)]$, called the transition matrix, which expresses the relationship

$$\{h(t)\} = [\Phi(t)]\{h(0)\} \tag{5.40}$$

for the complete period of $0 \leq t \leq T$. Floquet recognized that the content of the transformation matrix transcending one full period determines the solution throughout all periods. A way to obtain this transformation matrix is by integration in ref. [3] of the system equation as

$$\{h(T)\} = \int_{t=0}^{T} \{\dot{h}(t)\}\,dt = \int_{t=0}^{T} [A(t)]\{h(t)\}\,dt. \tag{5.41}$$

Unfortunately, this integral in most practical cases cannot be evaluated analytically; therefore a numerical integration approach is needed. We divide the full period into equidistant time steps as

$$\Delta t = \frac{T}{n}, \text{ with } t_i = i\Delta t, i = 1,..n.$$

Friedmann and Hammond in ref. [4] proposed the Gill variation of the fourth-order Runge–Kutta method, which is defined by the equation

$$\{h(t_{i+1})\} = \{h(t_i)\} + \frac{\Delta t}{6}\left(\{k_1\} + (2-\sqrt{2})k_2 + (2+\sqrt{2})k_3 + \{k_4\}\right), \qquad (5.42)$$

where the coefficients are

$$\{k_1\} = [A(t_i)]\{h(t_i)\},$$

$$\{k_2\} = \left[A\left(t_i + \frac{\Delta t}{2}\right)\right]\left(\{h(t_i)\} + \frac{\Delta t}{2}\{k_1\}\right),$$

$$\{k_3\} = \left[A\left(t_i + \frac{\Delta t}{2}\right)\right]\left(\{h(t_i)\} + \left(\frac{1}{\sqrt{2}} - \frac{1}{2}\right)\Delta t\{k_1\} + \left(1 - \frac{1}{\sqrt{2}}\right)\Delta t\{k_2\}\right),$$

$$\{k_4\} = [A(t_i + \Delta t)]\left(\{h(t_i)\} - \frac{1}{\sqrt{2}}\Delta t\{k_2\} + \left(1 + \frac{1}{\sqrt{2}}\right)\Delta t\{k_3\}\right).$$

Executing some of the posted operations and introducing intermediate matrices result in the following form of the coefficients:

$$\{k_2\} = \left[A\left(t_i + \frac{\Delta t}{2}\right)\right]\left([I] + \frac{\Delta t}{2}[A(t_i)]\right)\{h(t_i)\} = [A_2(t_i)]\{h(t_i)\},$$

$$\{k_3\} = \left[A\left(t_i + \frac{\Delta t}{2}\right)\right]\left([I] + \left(\frac{1}{\sqrt{2}} - \frac{1}{2}\right)\Delta t[A(t_i)] + \left(1 - \frac{1}{\sqrt{2}}\right)\Delta t[A_2(t_i)]\right)\{h(t_i)\}$$

$$= [A_3(t_i)]\{h(t_i)\},$$

$$\{k_4\} = [A(t_i + \Delta t)]\left([I] - \frac{1}{\sqrt{2}}\Delta t[A_2(t_i)] + \left(1 + \frac{1}{\sqrt{2}}\right)\Delta t[A_3(t_i)]\right)\{h(t_i)\} = [A_4(t_i)]\{h(t_i)\}.$$

The steps of the numerical integration will be of the form

$$\{h(t_{i+1})\} = [A_1(t_i)]\{h(t_i)\}, \qquad (5.43)$$

where the newly introduced matrix is

$$[A_1(t_i)] = [I] + \frac{\Delta t}{6}\Big([A(t_i)] + \big(2 - \sqrt{2}\big)[A_2(t_i)] + \big(2 + \sqrt{2}\big)[A_3(t_i)] + [A_4(t_i)]\Big). \qquad (5.44)$$

In detail the integration proceeds as

$$\{h(t_1)\} = [A_1(t_0)]\{h(t_0)\},$$

$$\{h(t_2)\} = [A_1(t_1)]\{h(t_1)\} = [A_1(t_1)][A_1(t_0)]\{h(t_0)\},$$

$$...$$

$$\{h(t_i)\} = [A_1(t_{i-1})][A_1(t_{i-2})]...[A_1(t_0)]\{h(t_0)\}.$$

This process is executed up to any time steps. For $i = n, t_n = T$, we obtain

$$\{h(T)\} = \prod_{i=1}^{n} [A_1(T - i\Delta t)]\{h(0)\}. \qquad (5.45)$$

Note the importance of the reverse order in the product computation. The transition matrix may be computed as

$$[\Phi(T)] = \prod_{i=1}^{n} [A_1(T - i\Delta t)], \qquad (5.46)$$

and following Equation (5.40) we can obtain the solution at the end of the period as

$$\{h(T)\} = [\Phi(T)]\{h(0)\}. \qquad (5.47)$$

The Floquet solution at the end of the period is

$$\{h(T)\} = [\Psi(T)][e^{\eta T}]. \qquad (5.48)$$

Equating the last two equations, exploiting that $\{h(0)\} = [\Psi(0)]$ and $[\Psi(T)] = [\Psi(0)]$, yields

$$[\Psi(0)]^{-1}[\Phi(T)][\Psi(0)] = [e^{\eta T}]. \qquad (5.49)$$

On the other hand, the spectral decomposition of the transition matrix may be written in the form

$$[\Psi]^{-1}[\Phi(T)][\Psi] = [\Lambda]. \tag{5.50}$$

Comparing the last two equations, we conclude that the eigenvectors of the transformation matrix provide the periodic Floquet eigenvectors of the system solution at the beginning of the period. The eigenvalues of the transition matrix, encapsulated in the diagonal matrix $[\Lambda]$, produce the exponential terms (Floquet eigenvalues) of the periodic solution. The corresponding terms of the diagonal matrices are equal:

$$\eta_j = \frac{1}{T} \ln(\lambda_j). \tag{5.51}$$

Because the eigenvalues are complex, we use Euler's identity, $\lambda_j = \xi_j + i\zeta_j = r_j e^{\theta j}$, as

$$\eta_j = \frac{\ln(r_j)}{2T} + i \frac{\tan^{-1}(\zeta_i/\xi_i)}{T} = \frac{\ln(r_j)}{2T} + \frac{i\theta_j}{T}. \tag{5.52}$$

The inverse tangent function used to obtain θ_j is multivalued; hence the above is true for any imaginary part obtained by adding any integer multiple of 2π to θ_j. However, fixing the arbitrary integer to a certain value renders the solution unique. The stability of the system is determined from the real part of the eigenvalue. Once the exponents and the periodic eigenvectors at the starting time are available, the periodic eigenvectors at any time are computed as

$$[\Psi(t)] = [\Phi(t)][\Psi(0)][e^{-\eta t}]. \tag{5.53}$$

These periodic eigenvectors represent all integer harmonics equally. The transition matrix at a certain time $t_j \le t \le t_{j+1}$ may be computed as

$$[\Phi(t)] = \prod_{i=1}^{j} [A_1(t - i\Delta t)]. \tag{5.54}$$

The methodology introduced so far in this section produced the general solution to the homogeneous differential equation that is used for the analysis of wind turbines.

The solution of the inhomogeneous system requires additional work. Let us consider the inhomogeneous problem of the form

$$\{\dot{h}(t)\} = [A(t)]\{h(t)\} + \{f(t)\}, \tag{5.55}$$

where the linear problem's load vector is formed as

$$\{f(t)\} = \left\{ \begin{array}{c} [M(t)]^{-1}\{F(t)\} \\ \{0\} \end{array} \right\}. \tag{5.56}$$

Let us assume that the particular solution of Equation (5.55) is

$$\{h_p(t)\} = [\Phi(t)]\{c\}, \tag{5.57}$$

a product of the transition matrix and a vector of constant coefficients. The variation of the constant parameters and differentiation leads us to

$$\{\dot{h}_p(t)\} = [\Phi(t)]\{\dot{c}(t)\} + [\dot{\Phi}(t)]\{c(t)\} = [\Phi(t)]\{\dot{c}(t)\} + [A](t)[\Phi(t)]\{c(t)\}. \tag{5.58}$$

On the other hand, substitution into Equation (5.55) produces

$$\{\dot{h}_p(t)\} = [A(t)]\{h_p(t)\} + \{f(t)\} = [A(t)][\Phi(t)]\{c(t)\} + \{f(t)\}. \tag{5.59}$$

Comparing the two solutions, we obtain the equation for the coefficients of a particular solution as $[\Phi(t)]\{\dot{c}(t)\} = \{f(t)\}$. Integration yields the coefficients as

$$\{c(t)\} = \int_{t=0}^{T} [\Phi(t)]^{-1}\{f(t)\}dt.$$

Finally, the general solution of the nonhomogeneous problem is the sum of the solution of the homogeneous problem and the particular solution:

$$\{h_g(t)\} = [\Psi(t)][e^{nt}] + [\Phi(t)] \int_{t=0}^{T} [\Phi(t)]^{-1}\{f(t)\} dt. \tag{5.60}$$

The practical results of this solution process will be demonstrated in Chapter 10, Section 10.4 by analyzing a wind turbine example.

Part II

Engineering Analysis of Rotating Structures

6

Resonances and Instabilities

The evaluation of the resonances and the stability of a rotating system is an important part of engineering dynamics. In particular, the Campbell diagram is the tool of visualizing the resonance points of a rotor. Because the Campbell diagram depends on whether the system is analyzed in the fixed or the rotating system, we first discuss some modeling considerations.

6.1 Analysis Type vs. Modeling Approach

In the prior discussions we saw that there are differences in the contents of certain matrices of the equilibrium equation, depending on the choice of reference system. For example, in the fixed system only the nodal rotations were used in connection with the rotational matrices. In the rotating system, the nodal translations also contributed to the rotational components. These differences have a direct relationship with the modeling approaches.

For example, fixed system analysis cannot be used for a model with solid elements because the solid elements in the finite element method have no nodal rotations; the degrees of freedom corresponding to the rotational nodal displacements do not exist.

This can be overcome by covering the solid elements with a thin layer of shell elements. This method, however, should be used with care. Generally, it is better to analyze this type of model in the rotating reference frame.

Models built from shell elements may be analyzed in both systems. However, care must still be taken when analyzing in the fixed reference system, because if the shell structure is thin walled, the sections will not remain planar. Furthermore, the shell normal degrees of freedom are not always defined for all shell elements in commercial software, or the rotations normal to the elements may only be approximations and there is no inertia in such models. Hence, it is best to analyze this type of structure also in the rotating system.

When the rotor structure is supported by bearings or by an elastic supporting structure, the coupled system developed in Chapter 2 must be the basis of analysis. In general, the resulting equations have periodic terms in the rotor angle. The effect of bearing is the subject of Section 6.6.

In addition to this, symmetry conditions must also be considered. A symmetric rotor with symmetric bearings or supporting structure can be analyzed in both systems. An unsymmetric rotor with symmetric bearings must be analyzed in the rotating system. A symmetric rotor in unsymmetric bearings must be analyzed in the fixed system if the uncoupled solution is used.

If the supporting structure is symmetric, the stiffness and damping of the bearings can be transformed to the rotating system and no periodic terms will arise. If the supporting structure is unsymmetric with respect to the axis of rotation, but the rotor is symmetric, the equation in the fixed system has no periodic terms and the analysis can be done in the fixed reference system, provided the rotor model is suitable for this type of analysis.

When an elastic rotor is mounted on an elastic fixed structure, the deformations and rotations on the point where the rotor is attached will turn in the rotating reference system. This leads to periodic terms in the equations of motion of the whole system. Such applications will be discussed in Chapter 10.

For many rotating machines, the foundation can be regarded as stiff, but the bearings are soft. A common type of bearing is the so-called journal bearing with oil film lubrication, discussed in detail in Chapter 8, Section 8.4. The stiffness and damping values of this type of bearing are unsymmetric, and the reduced-order solution needs complex modal reduction, as described in Chapter 4, Section 4.6. The conventional real modal reduction cannot be used in this case. The direct method may be used but requires large computational time.

6.2 Resonances and Instabilities

In the last chapter we saw the computational technique to solve the various complex eigenvalue problems of the common form

$$(\lambda^2[M] + \lambda[\tilde{D}] + [\tilde{K}])\{\varphi\} = \{0\}. \tag{6.1}$$

The content of the matrices was different in the fixed and the rotating reference systems and was dependent on the rotational speed; hence a solution was obtained for a range of selected rotor speeds. The solutions at each rotor speed are the complex conjugate pairs of eigenvalues:

$$\lambda = \delta \pm i\omega.$$

The real part is a measure of the amplitude amplification or decay. Positive values lead to an increase in amplitude with time, and the solution is unstable. The system is stable when the real part is negative. The damping of the eigenvalue is defined as

$$\zeta = \frac{\delta}{\omega}, \tag{6.2}$$

and it is the fraction of critical viscous damping. The imaginary part represents the oscillatory part of the solution. The natural frequency is

$$f = \frac{\omega}{2\pi} \quad [Hz]. \tag{6.3}$$

The Campbell diagram is obtained by plotting the frequencies over the rotor speed and connecting the appropriate solutions by lines. The diagram also contains straight lines called 1P, 2P, etc., representing the locations where the frequency and the rotational speed, or its various multiples, are equal. These lines will be instrumental in identifying the critical rotor speeds.

For different complex modes, the frequency may increase or decrease with rotor speed, and the frequency may couple with or cross those of other modes. It is therefore important to be able to draw lines that connect the correct solutions. This process is called mode tracking.

Mode tracking may be done in the following way. A loop is made over the rotor speed and then over the solutions. An extrapolation of the real and imaginary part is done in order to find the expected solution for the next rotor speed. Then the solutions for this speed are scanned, and the closest solution is selected. For some problems it is useful to reverse the order of the two loops. For complicated models this process may be troublesome, and some parameters may need to be applied.

For the points of intersection with the 1P line in the fixed system, the frequency is equal to the rotor speed: $\omega = \Omega$. To obtain these points, the following eigenvalue problem is solved:

$$(-\Omega^2[M] + i\Omega^2[C] + [\tilde{K}])\{\varphi\} = \{0\}. \tag{6.4}$$

Here the damping was neglected. The imaginary parts of the solutions are the critical rotor speeds. For the modes that don't cross the 1P line, the imaginary part is zero, and there are no critical speeds. This is also called synchronous analysis. A full analysis including the Campbell diagram should always be done in order to better interpret the results.

The critical speeds may depend on the direction of the rotation. In the rotating reference system the critical speed for the forward whirl direction is the crossing with the abscissa. Hence $\omega = 0$. The synchronous equation for this case is therefore

$$[\tilde{K}]\{\varphi\} = ([K] - \Omega^2([Z] - [K_G]))\{\varphi\} = \{0\}. \tag{6.5}$$

The resonance for the backward whirl motion is the crossing with the 2P line, and $\omega = 2\Omega$ is inserted into the eigenvalue problem of Equation (6.4). The synchronous equation is then

$$(-4\Omega^2[M] + 4i\Omega^2[C] + [\tilde{K}])\{\varphi\} = \{0\}. \tag{6.6}$$

We will use the simple rotating mass particle, the subject of our deriva-
tions in earlier chapters, to demonstrate the critical speed analysis. A rotating mass particle can be analyzed either in the rotating or in the fixed system;
hence it is a perfect example to demonstrate the specifics of the critical speed
analysis process in the two systems.

6.3 Critical Speed of Rotating Mass

Let us analyze a mass of 1 kg, attached to two springs of stiffness 1 N/m in
x- and y-directions. The mass is rotating about the z-axis. We execute the
analysis in the rotating system and will convert the results to the fixed system *a posteriori*. Accounting for the centrifugal softening and the gyroscopic
matrix, and plotting all solutions for both negative and positive rotor speeds,
the eigenfrequencies shown on Figure 6.1 are obtained.

The natural frequencies for motion in the x- and y-directions are 1 rad/s.
At a rotor speed of 1 rad/s the stiffness softening due to the centrifugal force

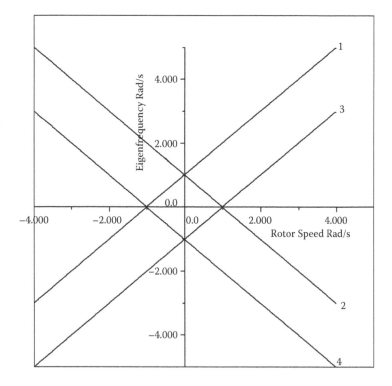

FIGURE 6.1
All solutions, positive and negative speeds.

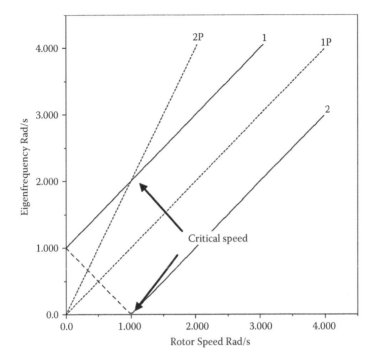

FIGURE 6.2
Campbell diagram for the rotating mass.

is $m\Omega^2 = 1\,N/m$ and is equal to the elastic stiffness. The first quadrant of the plot in Figure 6.1 is the Campbell diagram in Figure 6.2, showing only the positive eigenfrequencies as a function of the positive rotor speed.

In the rotating system the critical speed for the forward whirl mode is at the zero frequency point, and the crossing with the backward whirl is the crossing with the 2P line (two per revolution). The whirling motions of the mass point were calculated for the rotor speed of 0.5 rad/s based on the complex eigenvectors. The forward whirl is in the same sense as the rotor rotation and the backward whirl is in the opposite sense. The whirling directions are used in the Campbell diagram shown in Figure 6.2. The forward whirl solution is shown as solid lines, and the backward whirl solution is show as dashed lines. At the rotor speed of 1 rad/s, the backward whirl frequency goes to zero. Above this speed, the forward whirl motion of the solution with negative frequency on Figure 6.1 gets positive. The orbits of the whirling motions are shown in Figure 6.3.

The whirl directions can be used to convert the solutions from the rotating reference frame to the fixed system by adding and subtracting the rotor speed to the solutions. The converted solutions are shown in Figure 6.4. This result is equivalent to the results one would obtain by analyzing the mass point in the fixed system. In the fixed system there are no rotor

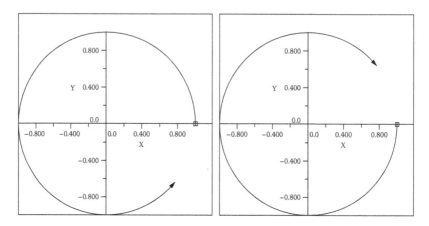

FIGURE 6.3
Orbit plots of the forward and backward whirl motions.

dynamic effects because there is no local rotation and no inertia moment of the rotating mass point; hence the solution will be a straight line as shown in the figure. In the fixed system shown in Figure 6.4 the critical speed is found at the crossing point between the mode line and the excitation line denoted by 1P.

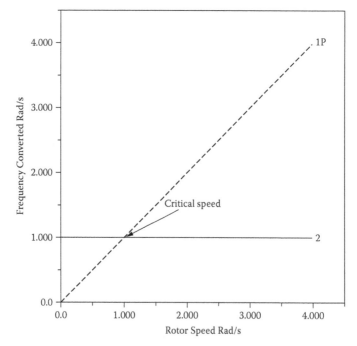

FIGURE 6.4
Solutions converted to the fixed system.

To consider the other topic of this chapter, that of stability, we turn to a model that is a little bit more complex and has theoretical solutions to validate the results.

6.4 The Laval Rotor

A steam turbine was built by Gustav de Laval around 1889 and was run up to 30,000 RPM. The theory of this rotor was developed by Henry Jeffcott around 1919. Hence the type of rotor model consisting of a rotating disk on a massless flexible shaft with bearings at the ends is now commonly called the Laval or Jeffcott rotor. This type of model is also called a line-model because all finite element nodes are located along the rotor axis, as shown in Figure 6.5.

The stiffness values of the bearings are equal in the x- and y-directions. This means that when considering the displacements and the forces, the stiffness values are invariant with respect to the reference system. Furthermore, they are also constant and nonperiodic in the rotating reference system. The rotor also expresses polar symmetry with respect to the rotor axis. Therefore this model can be analyzed in both systems.

FIGURE 6.5
Finite element model of a Laval rotor.

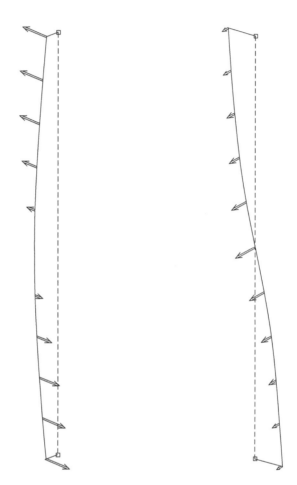

FIGURE 6.6
Bending modes of Laval rotor.

The mass of our Laval rotor is 40 kg, and the polar moment of inertia is 5.0 kg m^2. The rotor is attached to the midpoint of the shaft. The length of the shaft is 1000 mm and the diameter is 64.08 mm. The bearing stiffness is 3898 N/mm in both the x- and y-directions and is equal for both ends. This leads to a translational natural frequency of 50 Hz and a tilting frequency of 100 Hz. The first two bending modes are shown in Figure 6.6, where the arrows represent the modal rotation vector. The motion of the rotor mass is pure translation on the left, and the motion is a pure tilting of the disk on the right. The Campbell diagram in the fixed system is shown in Figure 6.7.

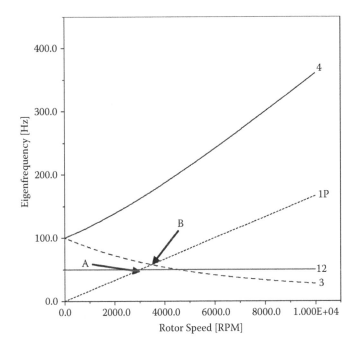

FIGURE 6.7
Campbell diagram of Laval rotor in the fixed system.

The translational modes 1 and 2 are straight lines as for the simple rotating mass. The critical speed for the translational modes is at 3000 RPM and is denoted by A in the figure. The curves denoted by 3 and 4 are the backward and forward tilting modes. There is no critical speed for the forward whirl mode, and for the backward whirl mode it is at 3464 RPM. This is point B in the figure. The same model is calculated in the rotating reference system in Figure 6.8

In this case the translational mode (1,2) behaves similarly to that of the rotating mass in the prior section. The critical speed for the forward translation mode is denoted with AF and is the point of zero frequency. This means that the whirling motion is synchronous with the rotor speed. The backward whirl resonance is denoted by AB and corresponds to the crossing with the 2P line. Both are at 3000 RPM as expected based on the fixed system results. The backward whirl resonance for the tilting mode is denoted by B, and again it is at the same 3464 RPM as in the fixed reference system case. There is no zero frequency point for this mode and therefore no resonance of the forward whirl tilting mode. In this case there is no coupling between the translation and tilting modes. If the rotor disk is not located in the midpoint of the shaft, coupling between translation and tilt would occur.

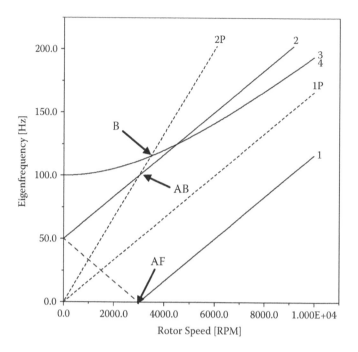

FIGURE 6.8
Campbell diagram in the rotating system.

6.5 Influence of Damping

As we saw in Chapter 3, damping is comprised of components acting on the rotating and nonrotating parts. Damping in the rotating part was called internal damping. Adding this damping to the Coriolis matrix, the total damping matrix becomes unsymmetric, and the real part of the solution becomes positive for speeds above the critical speed. This is an unstable scenario.

Damping in the bearings and the supporting structure, called the external damping, will stabilize the rotor, and the instability will occur at a higher speed. For the eigenvalue $\lambda = \delta + i\omega$ the damping is calculated as the ratio of the real and imaginary part of the solution:

$$\zeta = \frac{\delta}{\omega} \tag{6.7}$$

Because the eigenfrequencies are different for the rotating and fixed systems, the damping is also different. The damping in the fixed system for the Laval rotor model of the last section is shown in Figure 6.9. The critical speed occurs at 3000 RPM, where mode 2 rises above the horizontal axis. The computed damping is plotted for the rotating system in Figure 6.10.

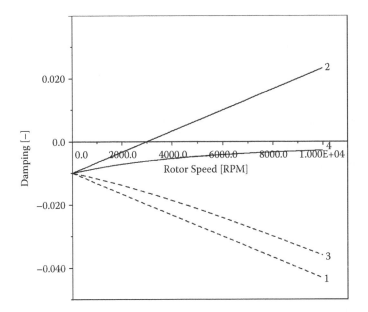

FIGURE 6.9
Damping calculated in the fixed system.

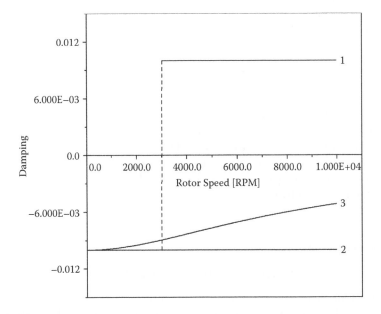

FIGURE 6.10
Damping calculated in the rotating system.

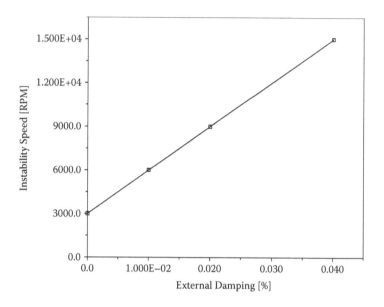

FIGURE 6.11
Instability speed as a function of external damping.

In the rotating system there is a jump in the damping curve at the critical speed. Note that the numbering of the modes is arbitrary, a fixture of the mode tracking procedure. The same modes are critical in both systems.

When external damping is increased in the bearings, the change of any instability can be calculated with the formula

$$\Omega_{Unstable} = \Omega_0 \left(1 + \frac{\zeta_E}{\zeta_I} \right),$$

where ζ_E, ζ_I are the external and internal damping, respectively. The change of a critical speed of a rotor with 1% internal damping, due to increasing the external damping, is shown in Figure 6.11.

Even a small amount of damping in the bearings leads to a higher speed of instability; hence this presents a tool to extend the stable operational range of the rotating structure.

6.6 Unsymmetric Effects of Bearing and Rotor

The effect of bearings is frequently unsymmetric in stiffness and damping. This is the case for the aforementioned journal bearings that will be discussed in detail in Chapter 8, Section 8.4. Also, the supporting structure can be

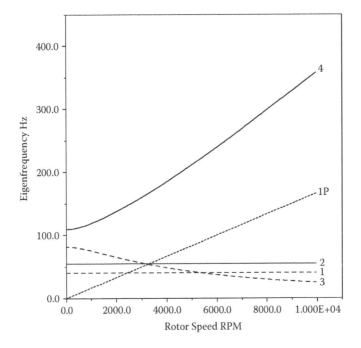

FIGURE 6.12
Campbell diagram of symmetric rotor in the fixed system.

unsymmetric, which means that the stiffness is different in x- and y-directions. In this case, the stiffness would contain periodic terms in the rotating system.

When the rotor is symmetric but the bearings are unsymmetric, the problem can be solved in the fixed system. Reducing the stiffness in the x-direction by a factor of 2 and increasing in the y-direction with a factor of 1.5 produces the result shown in Figure 6.12. Here the frequencies are different, and the forward and backward resonance speeds are different.

When the rotor is unsymmetric but the bearings are symmetric, the analysis must be done in the rotating system. Reducing the stiffness of the shaft by a factor of 2 in the x-direction and increasing by a factor of 1.5 in the y-direction, the eigenfrequencies of the structure change. The Campbell diagram is shown in Figure 6.13.

The translation mode now crosses the abscissa at two points. There are now two critical speeds equivalent to the two eigenfrequencies. The eigenfrequency (computed from the imaginary part) between the two critical speeds is zero and there are two real roots in this region. One of them is negative and is thus stable. The other one becomes positive, as shown in Figure 6.14. This means that the rotor is unstable between the two critical speeds. This instability is caused by the centrifugal force and is called centrifugal instability.

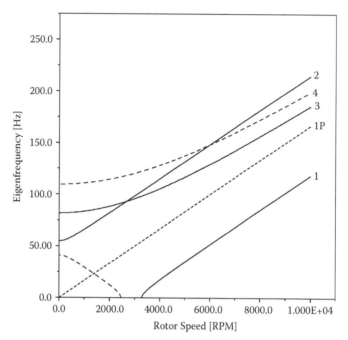

FIGURE 6.13
Campbell diagram of unsymmetric rotor in the rotating system.

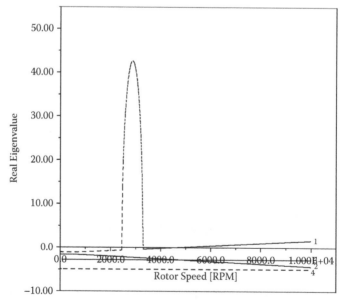

FIGURE 6.14
Real part of the solution.

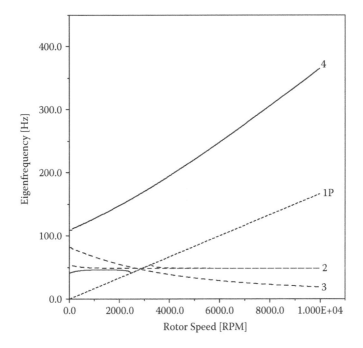

FIGURE 6.15
Converted to the fixed system.

The frequencies of the unsymmetric rotor converted to the fixed system are also shown in Figure 6.15. Note the corresponding regions of the instability in the two systems.

6.7 A Rotating Tube

After discussing two theoretical examples, let us consider a more practical example of a rotating cylinder with a thin wall, as shown in Figure 6.16.

The outer diameter of the cylinder is 0.3 m and the wall thickness is 0.02 m. The material of the cylinder is steel with Young's modulus of 2.1E+11 N/m² and a density of 7850 kg/m³. (This model is from ref [31]) The bearings are assumed to be stiff. Because this rotor has no inertia due to the modeling considerations explained in Section 6.1, it must be analyzed in the rotating system. The first three bending modes and the first shear mode are shown in Figure 6.17. These are so-called global modes.

Besides these modes, there is also a global torsion mode and several local modes with different lobe structures, as shown in Figure 6.18.

The Campbell diagram in the rotating system is shown in Figure 6.19. The bending modes show a similar behavior to the rotating mass and to

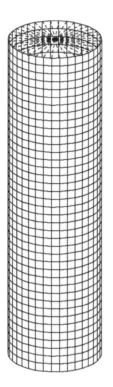

FIGURE 6.16
Rotating cylinder with a thin wall.

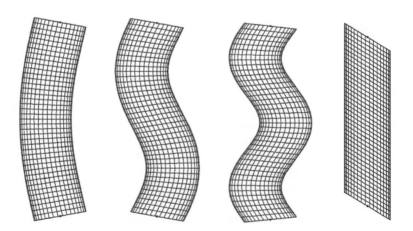

FIGURE 6.17
Bending and shear modes of the cylinder.

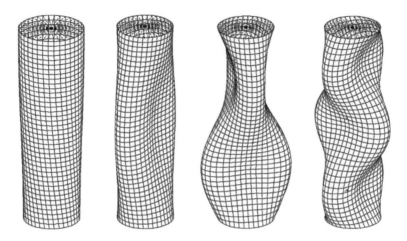

FIGURE 6.18
Torsion, and local modes.

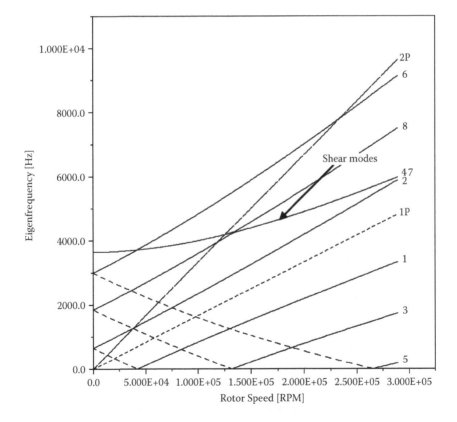

FIGURE 6.19
Rotating system Campbell diagram of rotating tube.

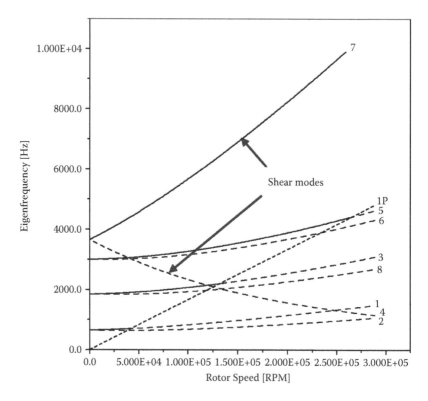

FIGURE 6.20
Fixed system results.

the translation modes of the Laval rotor described in the prior sections. The shear mode (number 7) looks similar to the tilting mode of the Laval rotor, denoted by 3 in Figure 6.8.

There is, however, a coupling between shear and bending. This can be seen in the plots of the higher bending modes in Figure 6.20 converted to the fixed reference system. The sections are not horizontal but tilted. Due to this tilting, the frequencies of the bending modes are also increasing with speed in the fixed system.

The above analysis results were obtained with the real modal method described in Chapter 4. This has the advantage that the order of the global finite element matrices is reduced to the number of modes extracted. The other advantage is that it is possible to select the modes to be accounted for in the rotor dynamic analysis. The disadvantage is that a sufficiently large number of real modes must be included in the modal space to avoid truncation errors, especially for the higher modes. In the shell model analyzed here, the first 100 modes were included in the modal space formulation.

As discussed in Section 6.1 there are different modeling possibilities for certain physical structures. This rotating tube structure can also be

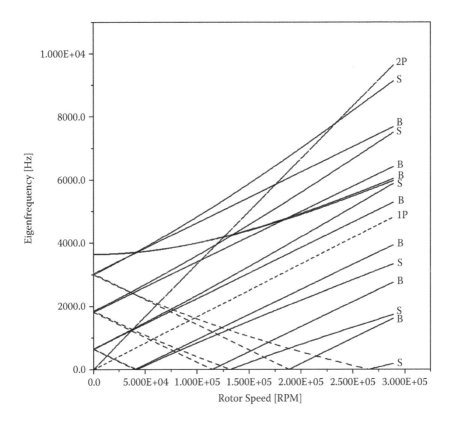

FIGURE 6.21
Comparison of the eigenfrequencies for the shell and the beam models.

approximated by hollow beam finite elements. In this case, the shear factor of 0.5 must be applied which is the factor between effective shear area and total area. The mass and inertia moments must also be entered for each node in order to calculate the shear modes.

The comparison of the two models is shown in Figure 6.21. The eigenfrequencies are in good agreement but the rotor dynamic behavior is different. Both approaches generate the first critical speed around 35,000 RPM. The higher critical speeds are found at a lower value for the beam model as opposed to the shell model.

The reason for the higher critical speeds of the shell model is due to the centrifugal forces acting on the shell elements. This leads to increased membrane stress which results in a higher geometric stiffness. This effect does not occur in the beam model where all nodes are along the rotor axis. This is also the cause of the increasing difference in critical speeds with higher rotor speed.

In the Campbell diagrams shown above we analyzed only eight rotational modes. As mentioned above, there are also torsion and various local modes.

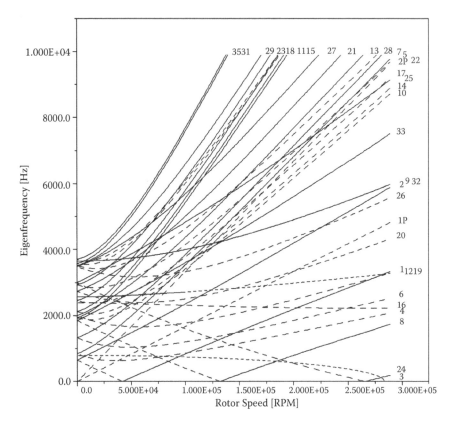

FIGURE 6.22
Campbell diagram of cylinder model with the first 35 modes.

Analyzing the first 35 rotational modes, the Campbell diagram in Figure 6.22 is obtained.

The radial and torsion modes are less affected by rotation and their eigenfrequencies remain almost constant. The local modes are all stiffened by the centrifugal force and they have no critical speeds, as demonstrated by their slopes being higher than the 1P line. Only the bending modes are going to zero at their respective critical speeds.

6.8 Rotating Model with Flexible Arms

Finally, let us consider a generic case, when the rotating component has arms, as a preparation for industrial rotors in later chapters. When the arms are flexible in bending, additional modes occur. The bending modes of the shaft will carry translation and tilting to the rotor, as shown in Figure 6.23.

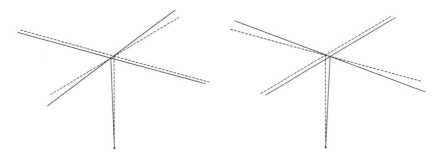

FIGURE 6.23
Shaft bending in x- and y-directions.

The shaft torsion results in the blade motion shown in Figure 6.24. These are also flap modes, where the flexible arms move in the z-direction, and lead-lag modes, where the arms move in a tangential direction. In this model the shaft is constrained at the attachment point at the bottom, and the motions of the arms can be classified as flap and lead-lag modes.

Figure 6.25 shows the symmetric and Figure 6.26 shows the antisymmetric flap modes. There are also lead-lag modes that can be symmetric, as shown in Figure 6.27, or antisymmetric, as shown in Figure 6.28, depending on the direction of motion and the phase between the arms.

These modes are, in principle, similar to those of an aircraft propeller or a wind turbine (topics of Chapters 9 and 10), although the dimensions and proportions are different. The above notations are mainly used for aircraft propellers. The lead-lag modes are called edgewise modes in the wind turbine industry.

An example flexible rotor analysis yielded the frequencies in Table 6.1 for the above mode shapes. The Campbell diagram for the same rotor is shown

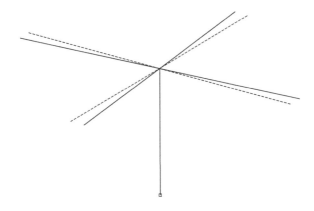

FIGURE 6.24
Shaft torsion.

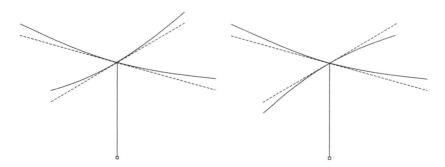

FIGURE 6.25
Symmetric flap modes, in-phase and anti-phase.

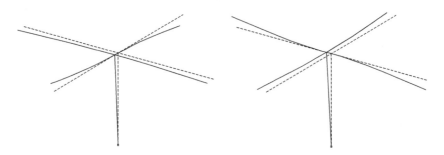

FIGURE 6.26
Antisymmetric flap modes.

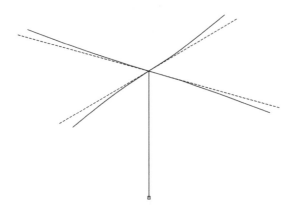

FIGURE 6.27
Symmetric lead-lag mode.

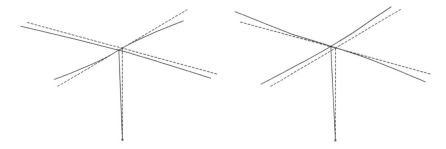

FIGURE 6.28
Antisymmetric lead-lag modes.

in Figure 6.29. The shaft torsion frequency shown as number 3 is constant and not influenced by the rotation. The two bending modes denoted by 1 and 2 show a similar behavior as translation modes of the simple models. This behavior is typical for helicopter rotors. The critical speed is at 5000 RPM.

The symmetric flap modes do not couple with other modes, as shown by lines 4 and 5. Their frequencies are increasing with speed due to the centrifugal force and the resulting geometric stiffness.

The symmetric lead-lag mode denoted by 8 also does not couple with other modes. It is specifically influenced by the geometric stiffness due to the centrifugal force shown in Figure 6.30. As shown in the figure, the dominant centrifugal force in the lead direction is $Z = \Omega^2 m u_y$. There is, however, a small motion in the x-direction because the blade length remains constant under the bending. The resulting centrifugal force is $Z = -\Omega^2 m u_x$, which produces a slightly decreasing eigenfrequency for the lead-lag motion. Due to these centrifugal components, the frequency increase is smaller for the lead-lag modes than for the flap modes.

TABLE 6.1

Eigenfrequencies and Modes

Mode Number	Eigenfrequency [Hz]	Symmetry	Notation
1	42.91	A	Shaft bending in x
2	42.91	A	Shaft bending in y
3	44.51	S	Shaft torsion
4	104.46	S	Flap
5	104.88	S	Flap
6	111.00	A	Flap
7	111.00	A	Flap
8	157.32	S	Lead-lag
9	250.41	A	Lead-lag
10	250.41	A	Lead-lag

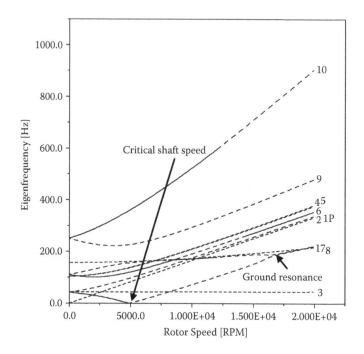

FIGURE 6.29
Campbell diagram in the rotating system.

As shown in Figure 6.29, in the antisymmetric flap modes (6 and 7) two blades are flap, but the other blades are moving in the rotor plane. Therefore they couple with the antisymmetric lead-lag modes denoted by 9 and 10. Around 17,000 RPM the shaft bending modes also couple with the blade lead-lag modes and the frequencies merge together. The real eigenvalues of the model are shown in Figure 6.31.

A small amount of damping was used in the model. Therefore there is a weak instability above the critical speed. This is normally prevented by an additional amount of damping in the fixed part of the structure.

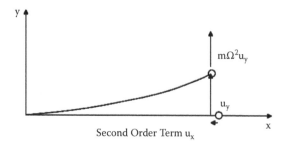

FIGURE 6.30
Centrifugal softening due to lead-lag motion.

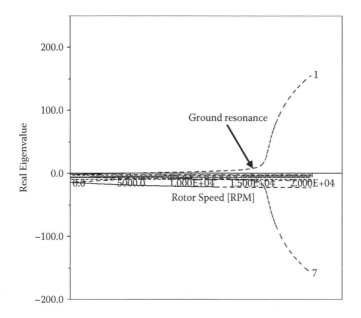

FIGURE 6.31
Real eigenvalues.

Around 16,000 RPM a strong instability occurs. The mode shape of the unstable solution is shown in Figure 6.32.

When the blades move in a lead-lag direction symmetrically, the center of mass of the two blades moves out of the rotor axis and gives rise to an additional centrifugal force that acts in the same direction as the rotor bending. This leads to a self-excitation of the rotor structure. It is therefore not a rotational resonance phenomenon but a mechanical instability. For helicopters

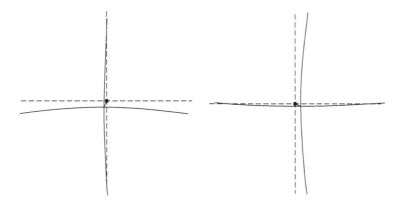

FIGURE 6.32
Complex mode of the stable solution at 17,500 RPM and 196.5 Hz.

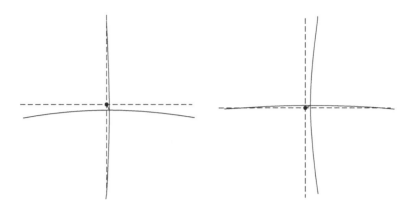

FIGURE 6.33
Complex mode of the stable solution at 17500 RPM and 196.5 Hz.

this phenomenon is called ground resonance and is discussed in more detail in Section 6.9. The corresponding stable solution is shown in Figure 6.33.

A common practice for helicopter and wind turbine blades is to add an angle, called a coning angle, to the blades. The name stems from the fact that the rotating blades now will form a cone, however slight, but not a plane. Adding a coning angle to the blade leads to a component in the radial direction for the flap motion, as shown in Figure 6.34. The additional centrifugal force due to the radial component is

$$Z = -\Omega^2 m u_r. \tag{6.8}$$

The centrifugal force therefore leads to a small negative stiffness.

As we saw in Chapter 1, the Coriolis force is perpendicular to the plane spanned by the axis of the rotation and the direction of the motion of the particle. Therefore the radial velocity leads to a Coriolis force in the direction perpendicular to the r-z plane. The Coriolis force coupling the flap and lag motion becomes

$$C = 2\Omega m \ddot{u}_r. \tag{6.9}$$

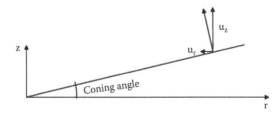

FIGURE 6.34
Coning angle of the blade.

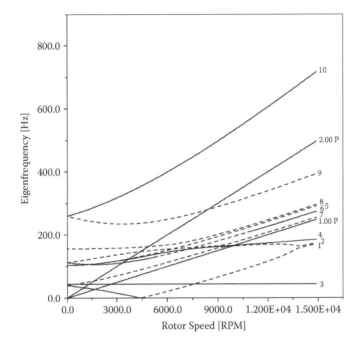

FIGURE 6.35
Campbell diagram for rotor with a coning angle.

The Campbell diagram incorporating the effect of the coning is shown in Figure 6.35. The results are similar to the previous case, but with a change in the critical speed location, a separation of the whirling directions of the flap modes, and coupling between flap and lag modes.

6.9 The Ground Resonance

In the early days of helicopter development, several helicopters were destroyed due to ground resonance problems. There have also been accidents when helicopters were landing on soft structures. Ground resonance was also observed on some helicopters shortly before take-off because the landing gear stiffness and damping are dependent on the load. Hinge-less helicopter blades are elastic in a flap and lag direction. Helicopters with hinges have the flap and lag hinge. The lead-lag motion of the blades couples with the translation motion of the mass of the fixed structure, resulting in the ground resonance.

The problem was first studied by Coleman and Feingold in the period from 1942 to 1947. The work was published in 1957 in ref. [5]. Their analytical blade model is shown in Figure 6.36. The pylon mass is denoted by m_f and is

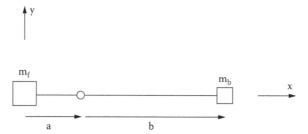

FIGURE 6.36
Helicopter blade with lead-lag bearing mounted on a pylon.

supported by springs in x- and y-directions. The blade mass is denoted by m_b and is located at the midpoint of the blade. The lag bearing is located at the radial distance from the rotor axis, and the distance from the bearing to the blade center of mass is b. The radius of gyration of the blade is r and there are n blades. Coleman and Feingold used dimensionless parameters to describe the geometry, lag stiffness, and mass parameters as

$$\Lambda_1 = \frac{a}{b\left(1+\frac{r^2}{b^2}\right)}, \ \Lambda_2 = \frac{K_\beta}{I}, \ \Lambda_3 = \frac{\mu}{2\left(1+\frac{r^2}{b^2}\right)}, \tag{6.10}$$

where the stiffness of the lead-lag spring at the lag hinge is denoted by K_β and the mass ratio is

$$\mu = \frac{n\,m_b}{m_f + n\,m_b}. \tag{6.11}$$

The inertia of a uniform cross-sectioned blade about its center of mass is

$$I_b = \rho A \frac{8b^3}{12} = \frac{1}{3}b^2\, m_b. \tag{6.12}$$

With a blade mass of $m_b = 2\rho Ab$, where A is the cross-section area, the radius of gyration becomes

$$r = \sqrt{\frac{I_b}{m_b}} = \frac{b}{\sqrt{3}}. \tag{6.13}$$

Substitution leads to the following expression for the mass ratio:

$$\mu = \frac{8\Lambda_3}{3}. \tag{6.14}$$

Coleman and Feingold introduced a reference eigenfrequency, computed as

$$\omega_r = \sqrt{\frac{K_x}{M}}, \tag{6.15}$$

where K_x and K_y are the hub stiffness in x- and y-directions, and $M = m_f + n\, m_b$ is the total mass of hub and blades. Only the x-direction term appears because the analytic model was symmetric. They also defined a dimensionless rotation speed in terms of the reference frequency as

$$\Omega_r = \frac{\Omega}{\omega_r}. \tag{6.16}$$

With these, the analytic solution ω_a is obtained from the equation

$$2\Lambda_3\left(\Omega_r^2 + \omega_a^2 - 1\right) + \left[4\Omega_r^2\omega_a^2 - \left(\Omega_r^2 + \omega_a^2 - 1\right)^2\right]$$

$$\left[-2\Lambda_3\left(\Omega_r^2 + \omega_a^2 + 1\right) + \omega_a^2 - \Lambda_2 - \Lambda_1\Omega_r^2\right] = 0. \tag{6.17}$$

We now recreate this analytical model with a finite element model with two blades, as shown in Figure 6.37. The stiffness of the pylon is equal in x- and y-directions. We chose a blade length of $b = 2.0$ m and a blade mass of $m_b = 30$ kg. The other data can be derived by selecting the above parameters. We selected $\Lambda_1 = 0.05$, $\Lambda_2 = 0.2$, and $\Lambda_3 = 0.1$, leading to $a = 0.13333$ m and the mass ratio of

$$\mu = \frac{8}{3}\, 0.1 = 0.26667.$$

In turn, this leads to the pylon mass

$$m_f = n\, m_b \frac{(1 - 26667)}{0.26667} = 2.75\, n\, m_b.$$

FIGURE 6.37
Finite element model of the rotor with two blades.

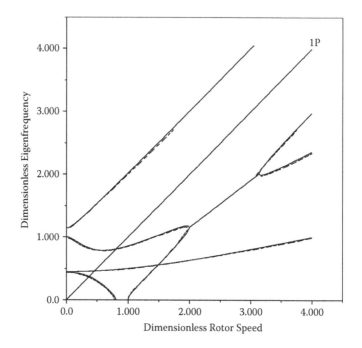

FIGURE 6.38
Campbell diagram of the two-bladed rotor.

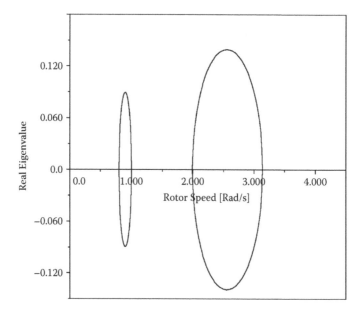

FIGURE 6.39
Real eigenvalues of the two-bladed rotor.

With two blades the mass is $m_f = 165.0$ kg. The moment of inertia of the blade about the lag hinge is computed (using $m_b = 30$ kg) as

$$I = b^2 m_b + I_b = b^2 m_b + \frac{1}{3}b^2\ m_b = b^2 m_b \frac{4}{3} = 160\ kg\ m^2.$$

With the inertia and the chosen $\Lambda_2 = 0.2$, the stiffness of the lead-lag spring at the lag hinge becomes

$$K_\beta = \Lambda_2\ I = 0.2 \times 160 = 32.0.$$

The model is analyzed in the rotating system. The results are shown as solid lines in Figure 6.38. The results of the analytical model are included as dashed lines. The agreement between the analytical model and the finite element computational model is very good.

There are two critical speeds at the dimensionless speed of 0.8 and 1.0, shown by the intersection of the horizontal axis (due to the analysis being in the rotating system). In this region the rotor is unstable, as shown in the plot of the real eigenvalues in Figure 6.39 by the smaller ellipse on the left. This is the centrifugal instability as encountered in Section 6.6.

The eigenfrequencies of the pylon and the blades merge between the speeds 2.0 and 3.13, as shown in Figure 6.38. In this region a strong instability can be seen in the plot of the real eigenvalues of Figure 6.39, demonstrated by the right-hand-side ellipse with the larger major axis. This is the ground resonance instability.

This comparison between the analytical and the finite element model serves two purposes. First, we explained the somewhat strange observation in the Campbell diagram in simple analytic terms. Second, we validated that the finite element method-based rotor dynamics, the subject of this book, are also correct in the most intriguing rotational scenarios.

After calculating resonances and stability, in practical circumstances response analysis must also be executed for various industry-dependent excitation types, which is discussed in the next chapter.

7

Dynamic Response Analysis

For the response analysis, different excitation forces need to be included. The most common excitation is the mass unbalance. For propellers and helicopter rotors the aerodynamic excitation is important. For wind turbines the gravity excitation must be included. This is also the case for horizontally mounted turbines and generators.

The critical speed is found at the crossing of the eigenfrequency line with the excitation line. A mass unbalance will excite the rotor in the forward whirl mode. For many applications, this is the critical speed. For other excitations the critical speed can be different.

7.1 Frequency Response without Rotation

The equation of motion of a single degree of freedom system, based on Equation (4.11), but replacing the matrices by scalar values and the generalized displacement vectors with the single displacement component of the particle, Chapter 4, is

$$m\ddot{u}(t) + d\dot{u}(t) + ku(t) = f(t). \tag{7.1}$$

Following the steps in Chapter 4, Section 4.2, in frequency response analysis the forces are harmonic with the form

$$f(t) = f(\omega)e^{i\omega t}. \tag{7.2}$$

The steady state solution is then also harmonic and presented by the form

$$u(t) = u(\omega)e^{i\omega t}. \tag{7.3}$$

Substituting these leads to the frequency response equation of the single mass point as

$$(-\omega^2 m + i\omega d + k)u = f, \tag{7.4}$$

where ω is the excitation frequency, and the solution and right-hand-side dependence on it is omitted for simplicity. The lone eigenvalue of the problem in the undamped case is

$$\omega_0^2 = \frac{k}{m}. \tag{7.5}$$

The physical damping is defined in terms of this eigenvalue as

$$d = 2\zeta\omega_0 m, \tag{7.6}$$

where the ζ is the so-called modal damping coefficient. With these substitutions the equation of motion can be written as

$$m\left(-\omega^2 + i2\omega\omega_0\zeta + \omega_0^2\right)u = f. \tag{7.7}$$

This damped equation is now solved for different excitation frequencies, producing the complex displacement solution of the form

$$u = \frac{f}{m}\frac{1}{-\omega^2 + i2\omega\omega_0\zeta + \omega_0^2} = u_{re} + iu_{im}, \tag{7.8}$$

with real and imaginary parts of

$$u_{re} = \frac{f}{m}\frac{\left(\omega_0^2 - \omega^2\right)}{\left(\omega_0^2 - \omega^2\right)^2 + \left(2\omega\omega_0\zeta\right)^2} \tag{7.9}$$

and

$$u_{im} = -\frac{f}{m}\frac{2\omega\omega_0\zeta}{\left(\omega_0^2 - \omega^2\right)^2 + \left(2\omega\omega_0\zeta\right)^2}. \tag{7.10}$$

The magnitude of the complex displacement solution of the mass particle is

$$|u| = \frac{f}{m} \frac{\sqrt{\left(\omega_0^2 - \omega^2\right)^2 + \left(2\omega\omega_0\zeta\right)^2}}{\left(\omega_0^2 - \omega^2\right)^2 + \left(2\omega\omega_0\zeta\right)^2}. \tag{7.11}$$

Let us now assume a single particle with mass equal to 1 and stiffness 39.4785 or $(2\pi)^2$. Because the general relationship is

$$\omega = 2\pi f$$

and

$$\omega_0 = \sqrt{\frac{k}{m}} = 2\pi,$$

the eigenfrequency is 1 Hz. We use a modal damping coefficient of 0.035 or 3.5%, producing a physical damping of $d = 0.4398$. Finally, let the excitation force also be 39.4785, and the resulting static deformation will be equal to 1. The solution is plotted in Figure 7.1.

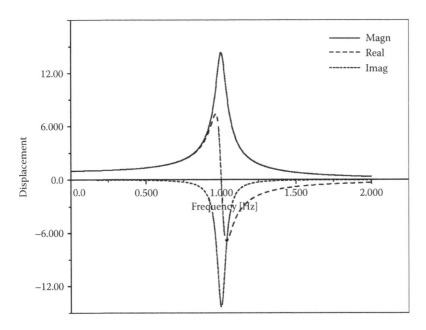

FIGURE 7.1
Displacement solution.

The real part of the solution is 1.0 at the zero frequency and 0.0 at the resonance frequency (1 Hz). The imaginary part is 0.0 at the zero frequency
and the response is real as in the static case. At the resonance frequency
the response is purely imaginary with a negative peak. The magnitude is
identical to the real part at 0 Hz and to the imaginary part (except for the
sign) at 1 Hz. The peak magnitude at the resonance point may be computed from Equation (7.9) by substituting the resonance frequency as

$$|u|_{max} = \frac{f}{m}\frac{1}{2\omega_0\zeta},$$
(7.12)

that with the given values yields 14.28.

If the damping is negative, the real part is unchanged but the imaginary
part changes sign. Therefore there would also be a peak for an unstable solution; hence the stability cannot be determined from a frequency response
analysis alone.

Plotting the imaginary part of the displacement solution against the real
part at equidistant frequencies, we obtain the Nyquist diagram shown in
Figure 7.2. In the diagram the calculated points are marked with the square

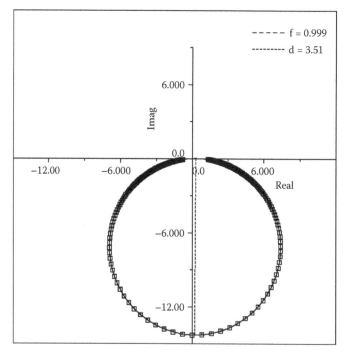

FIGURE 7.2
Nyquist plot of the response peak.

symbols. The resonance eigenfrequency is where the distance between the points is largest. Hence, taking the first derivative of the distance values along the circle, the eigenvalue can be determined. Here the result is approximately 1 Hz as expected.

7.2 Frequency Response with Rotation

Following Chapter 4, Equation (4.12), the equation in the fixed system is

$$\left(-\omega^2[M] + i\omega([D] + \Omega[C]) + ([K] + \Omega[K_D])\right)\{g\} = \{f\}, \tag{7.13}$$

with the additional terms of the Coriolis matrix and the structural stiffness damping term added to the classical finite element matrices. In the rotating system equation in addition to the terms in Equation 7.13, the geometric stiffness matrix and the centrifugal matrix also occur:

$$\left(-\omega^2[M] + i\omega([D] + 2\Omega[C]) + ([K] - \Omega^2[Z] + \Omega^2[K_G] + \Omega[K_D])\right)\{g\} = \{f\}. \tag{7.14}$$

Let us now analyze the last section's mass point frequency response with rotation. This model can be analyzed in both the fixed and the rotating systems. The Campbell diagram for the rotating system was shown in Chapter 6, Figure 6.2 and for the fixed system was shown in Chapter 6 Figure 6.4 already in Section 6.3. A modal damping coefficient of 2% was applied in the rotating part and 4% in the bearing in order to get a stable solution well above the critical speed.

A rotating excitation force may excite the forward and the backward whirl modes separately. A rotating force can be defined by multiplying the force in the x-direction with $\sin\Omega t$ and the force in the y-direction with $\cos\Omega t$ for the forward whirl, and $\sin\Omega t$, $-\cos\Omega t$ for the backward excitation.

In particular, the mass unbalance is a radial rotating force acting in synchrony with the sense of rotation. Such force will excite the forward whirl but not the backward whirl. The mass unbalance is proportional to Ω^2 and is modeled as a small mass located at a specific radial distance from the axis of rotation.

The analysis can be done in different ways:

1. Asynchronous analysis with constant rotor speed and increasing excitation frequency. This case is represented by a vertical line in the Campbell diagram.

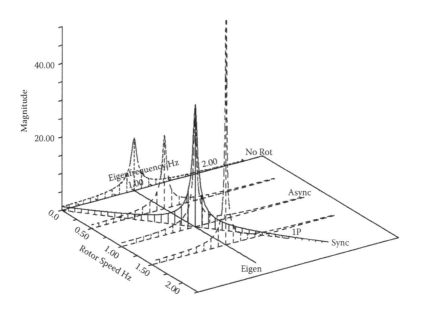

FIGURE 7.3
Response magnitudes plotted over the Campbell diagram for the forward whirl.

2. Synchronous analysis by keeping the excitation frequency proportional to the rotor speed. This case of the analysis is done along an excitation line like the 1P line.

3. It would also be possible to keep the excitation frequency constant and vary the rotor speed. This would be equivalent to a horizontal line in the Campbell diagram.

Figure 7.3 presents the Campbell diagram for the rotating mass point, already shown in Chapter 6, Figure 6.4, now visible in the horizontal plane for an analysis in the fixed system. The curve over the eigenfrequency axis represents the magnitude due to a force without rotation ($\Omega = 0$). The curves perpendicular to the rotor speed axis represent the response magnitudes for rotor speeds of 0.5, 1.0, and 1.5 Hz, respectively. The force is constant and does not increase with speed, as would be the case for a mass balance excitation. The resonance frequency is equal to the eigenvalue in the Campbell diagram.

The solid curve over the Sync line is the response magnitude of a synchronous analysis—case 2 in the above classification. At the point where the excitation line denoted by 1P is crossing the eigenfrequency, resonance occurs. The amplitude there is equal to the asynchronous analysis at 1-Hz speed. The peak, however, is broader because the angle between the 1P line and the eigenfrequency is smaller (here 45 degrees) than the crossing of the asynchronous excitation at 90 degrees. A similar plot for a force rotating backward is shown in Figure 7.4.

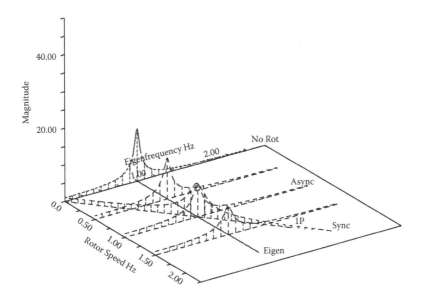

FIGURE 7.4
Response magnitudes for the backward whirl.

The reason for the increasing amplitudes in Figure 7.3 and the decreasing amplitudes in Figure 7.4 is in the shape of the damping curves shown in Figure 7.5. The damping curve corresponding to the forward whirl (2) is increasing, indicating decreasing damping of the mode as the rotational speed is increasing. On the converse, the damping curve of the backward whirl (1) is decreasing, which implies increased damping value.

The same results presented in the rotating system are shown in Figure 7.6 for the forward whirl case. Here the Campbell diagram of Chapter 6, Figure 6.2 is visible on the horizontal plane. The curve over the eigenfrequency axis is the response without rotation. The curves perpendicular to the rotor speed axis are the magnitudes for asynchronous analyses at 0.5-, 1.0-, and 1.5-Hz rotor speeds, respectively. At 0.5 Hz the response peak is located at 0.5-Hz frequency, and for 1.0-Hz rotor speed the peak is located at 0 Hz, which means a static deformation. Above this critical speed there are no more resonance peaks with the forward whirl.

The curve over the rotor speed axis is the synchronous response with forward excitation. The peak occurs at the critical speed at 1.0 Hz and the peak is as large as that for the asynchronous analysis at this rotor speed.

The results of response analyses with a backward-whirling force are shown in Figure 7.7. For the asynchronous response at 0.5 Hz there is a peak at the upper branch at 1.5 Hz. At the rotor speed of 1.0 Hz there is a peak at 0-Hz frequency and at the upper branch at 2.0 Hz. At 1.5 Hz there are two peaks because both branches in the Campbell diagram represent backward whirl motion.

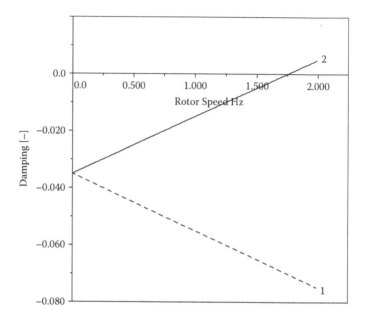

FIGURE 7.5
Damping of the forward whirl (2) and the backward whirl (1).

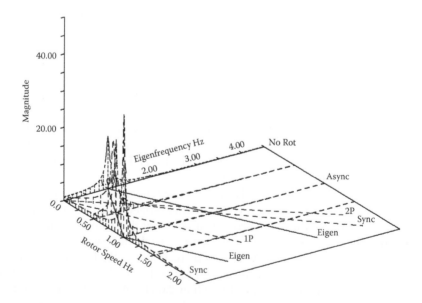

FIGURE 7.6
Response magnitudes for the forward whirl in the rotating system.

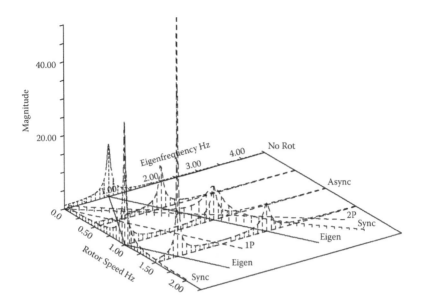

FIGURE 7.7
Response magnitudes for the backward whirl in the rotating system.

The curve along the rotor speed axis is the response to a static force and its peak is equal to the peak in Figure 7.6. At the critical speed of 1 Hz, the forward whirl motion ends and the backward motion is coming up. The curve along the 2P line has the maximum at the resonance point of the crossing of the 2P line with the eigenfrequency and has the same peak value as the curve of the asynchronous analysis at that speed. The peak is, however, broader due to the smaller angle between the curves, as was the case in the fixed system.

7.3 Transient Response without Rotation

In the transient analysis the equation for a single mass point, as shown in Equation (7.1), is solved by numerical integration. The initial conditions of the displacement and the velocity are zero. Therefore the response values will start at zero and the amplitudes will increase with time until the steady state condition is reached.

The transient response result for the same model as that used in the previous sections is shown in Figure 7.8. The mass point is excited at the 1-Hz

Computational Techniques of Rotor Dynamics

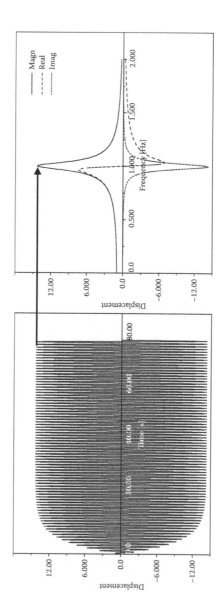

FIGURE 7.8
Magnitude of frequency response is equal to the steady state amplitudes of the transient response.

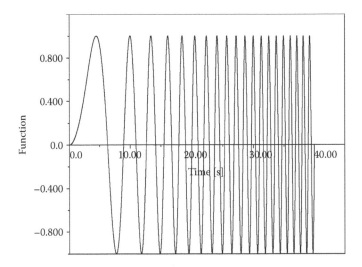

FIGURE 7.9
Linear sweep function.

eigenfrequency. The maximum amplitudes are equal to the values found in the frequency response analysis, and it takes about 20 seconds to reach the steady state condition.

The excitation function can be arbitrary, not necessarily harmonic, in the transient response analysis. Let us consider an excitation function such as that shown in Figure 7.9 with a gradually increasing frequency as visible from its profile. If the frequency of the function is linearly increasing as shown in the upper part of Figure 7.10, it is commonly called a linear sweep function. Depending on the rate of the frequency increase, it may be called a slow or a fast sweep. Such excitation functions may be obtained by trigonometric functions, specifically sine and cosine.

Let us now apply this excitation function to our model. The response to a slow sweep is shown in the lower part of Figure 7.10. Drawing a vertical line from the time of maximum amplitude to the sweep function, the frequency at that time can be found by drawing a horizontal line to the ordinate axis. The resonance is found at the frequency of 1 Hz, as expected.

The response to a fast sweep function is shown in Figure 7.11. Here the amplitudes are smaller and the peak occurs at a higher frequency because the structure does not reach the steady state condition. Note the time scale difference between the two diagrams; the resonance of the fast sweep is at 40.0 s, while that of the slow sweep is at 400.0 s.

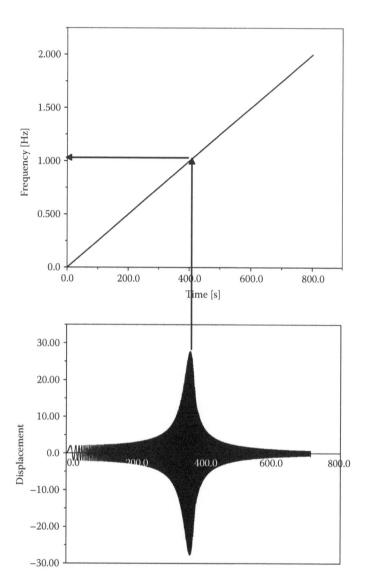

FIGURE 7.10
Resonance due to a slow sweep excitation.

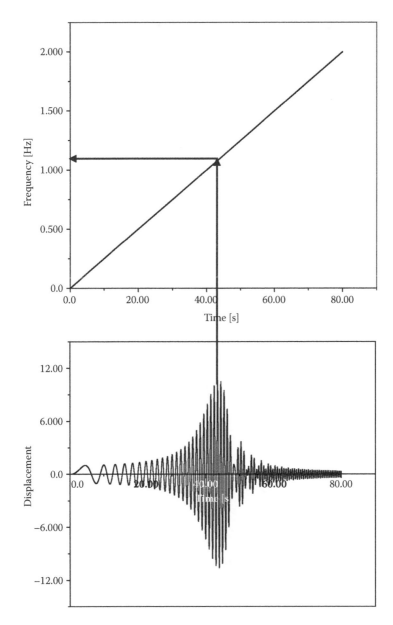

FIGURE 7.11
Resonance due to a fast sweep function.

7.4 Transient Response with Rotation

Let us recall the equations of motion, from Chapter 3, Equations (3.40) and
(3.41), for the fixed system as

$$[M]\{\ddot{g}(t)\} + ([D] + \Omega[C])\{\dot{g}(t)\} + ([K] + \Omega[K_D])\{g(t)\} = \{f(t)\}$$

and for the rotating system as

$$[M]\{\ddot{g}(t)\} + ([D] + 2\Omega[C])\{\dot{g}(t)\} + ([K] - \Omega^2[Z] + \Omega^2[K_G] + \Omega[K_D])\{g(t)\} = \{f(t)\}.$$

The excitation force is a function of time. A whirling excitation force can be
defined by two simultaneously applied sweeps of sine and cosine functions,
acting in x- and y-directions, respectively. Both synchronous and asynchronous analyses can be done in this case.

The motion of the mass point is now a spiral whirl motion, as shown in
Figure 7.12. The amplitude starts at zero and increases to the steady state
response, which is a circle with direction of motion in the sense of rotation.
The radius of the circle is the magnitude of the displacement of the particle:

$$|u| = \sqrt{u_x^2 + u_y^2}.$$

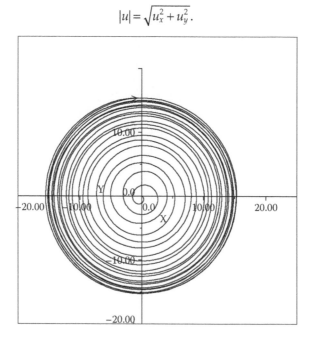

FIGURE 7.12
Orbit motion of the forward whirl in the beginning of the simulation.

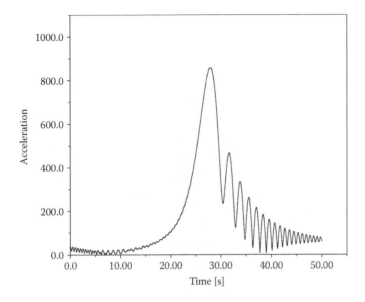

FIGURE 7.13
Magnitude of acceleration.

Similarly, the magnitude of the acceleration is of practical importance. An example is shown in Figure 7.13 for a fast sweep.

The responses to a slow sweep in synchronous and asynchronous scenarios are shown in Figure 7.14 and Figure 7.15, respectively. The corresponding displacement magnitudes, representing the envelopes of the response curves, are in Figures 7.16 and 7.17.

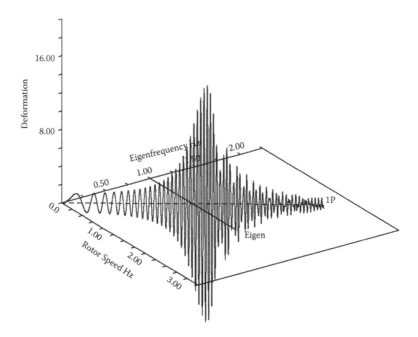

FIGURE 7.14
Synchronous analysis response.

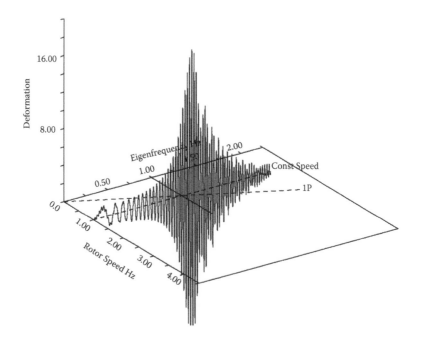

FIGURE 7.15
Asynchronous analysis response at 1-Hz rotor speed.

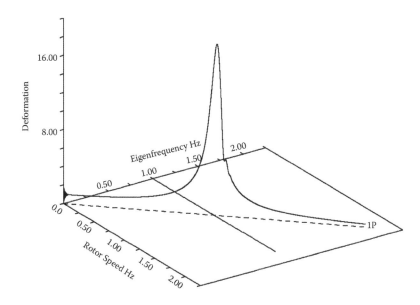

FIGURE 7.16
Magnitude of synchronous analysis.

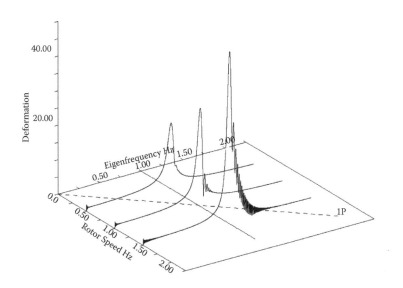

FIGURE 7.17
Magnitude of asynchronous analysis.

8

A Finite Element Case Study

The last two chapters described the industrially important stability and critical speed analyses, as well as the dynamic response computations, in connection with simplified rotor models. In this chapter we turn our attention to an industrial finite element model.

The model is a single turbine stage comprised of shaft, wheel, and blades and will be used to demonstrate the computational complexity and engineering time requirement for solving practical problems.

8.1 Turbine Wheel with Shaft and Blades

The stability and critical speed analysis of a turbine wheel with shaft and blades is a frequent application in the industry. Our example structure consisted of 30 blades attached to the wheel as shown in Figure 8.1. Note that the geometric components of the rotating structure are comprised of both simple cylindrical components (the shaft and the wheel) and special double curved surfaces (of the blades).

The geometric model was discretized with 45,350 node points and 24,199 mainly tetrahedral finite elements. The finite element model is shown in Figure 8.2. Besides the tetrahedron elements, the finite element model also contains four springs and dampers to present a simple bearing of the rotating turbine. The topic of industrial bearings will be discussed in more detail in Section 8.4. Because the model was dominated by solid elements, fixed case analysis was not appropriate and rotating system analysis was used.

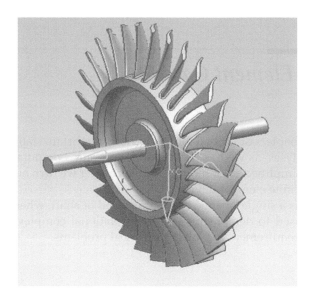

FIGURE 8.1
Turbine geometric model.

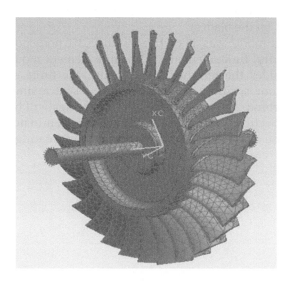

FIGURE 8.2
Turbine finite element model.

8.2 Engineering Analysis

The rotor dynamic analysis of the structure encompassed the range of 0 to 25,000 RPM with 100 intermediate steps. The Campbell diagram of the turbine, with visibly more content than in the earlier examples, is shown in Figure 8.3.

Several resonance speeds were found. The first at 4085 RPM is a forward whirl resonance. Backward whirl resonances (intersections with the 2P line) were found at 1389 RPM, at 4664 RPM, and at 16,969 RPM. The model also demonstrated instability at 4460 RPM, by intersecting with the 0P line (horizontal axis).

Inspecting the mode shapes, several interesting phenomena may be recognized at 20,000 RPM. Figure 8.4 shows the real (left) and imaginary (right) parts of a complex rotor disk tilting mode shape observed at 340.5 Hz. The lighter shaded areas are the higher deformations. It is very interesting to note the 90-degree phase angle difference between the real and the imaginary parts.

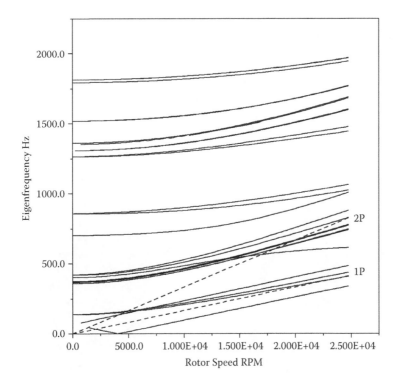

FIGURE 8.3
Turbine Campbell diagram.

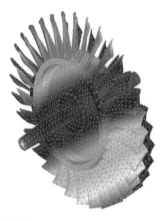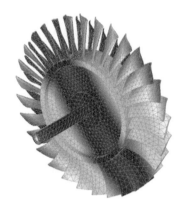

FIGURE 8.4
Turbine wheel tilting mode.

A disk axial displacement and the accompanying blade bending mode are found at 590 Hz, shown in Figure 8.5 again in the real (left) and imaginary (right) parts. The lighter shaded areas show the wheel translational motion on the left-hand side. The same mode results in all the blades moving in unison. Their difference is in the phase angle. An animation of this mode shape would show the wheel and blades' repetitive bending motion in time.

An example of the higher frequency local modes related to the blades, shown in the bundle of curves in the upper part of the Campbell diagram, is shown in Figure 8.6. These modes could be excited by the air flow but is beyond the scope of this book.

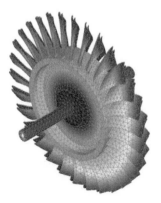

FIGURE 8.5
Axial displacement and global blade bending.

FIGURE 8.6
Local blade bending mode.

The application of the forces from the flow will be discussed in Section 8.5. Next we assess the computational costs of the above analysis.

8.3 Computational Statistics

The global finite element degrees of freedom were six times the number of node points, or 272,100. The boundary conditions applied to the bearing node points reduced this to 272,076 (partition k in Chapter 3, Section 3.6). Because the elements were tetrahedral, the rotational degrees of freedom were removed from the model, ultimately resulting in 136,035 free degrees of freedom (partition f in Chapter 3, Section 3.6). This fact rendered the use of fixed system analysis inapplicable for this case. Such reduction steps are standard in the finite element analysis of any structure, including rotational structures.

Two distinct computational solution approaches were executed to show their relative computational complexity. They were the direct free vibration solution of Chapter 4, Section 4.3 and the dynamic reduction version of Chapter 4, Section 4.6. The methods solve different-sized complex eigenvalue problems that are the most crucial components of the solution executed at each rotational speed. These complex eigenvalue analyses are the dominant factor in the performance of the analysis.

The direct free vibration solution executed the complex eigenvalue analysis at the problem size of 136,035. The execution times of the components of the eigensolution (discussed in detail in Chapter 5) on a standard workstation, in seconds, were

Matrix factorization: 399.0 s

Linear system solution: 2.7 s

Orthogonalization: 0.5 s

Eigenvalue solution: 1.6 s

Eigenvector computation: 75.3 s

The total CPU time of one eigensolution was almost 677.5 seconds; however, the collateral I/O requirements of these steps resulted in an elapsed time of 15 minutes and 14 seconds, or about a quarter of an hour. Because these steps were executed at each rotational speed, in this case 100 times, and there are other operations between the eigenvalue solutions, the overall time of the direct analysis was several days. Such a long analysis time does not fit into the usual engineering workflow; hence the justification for the various reduction methods introduced in Chapter 4.

One can certainly reduce this time by using fewer operational steps, and this is a feasible solution in the preliminary stages of the development of a new turbine. But it is more practical to execute such analyses with the desired number of rotational speeds by using one of the more advanced computational solutions, such as the various reduced-order solutions of Chapter 4.

To use the dynamic reduction method, the wheel and blades of the turbine, shown in Figure 8.7, were reduced to the boundary with the shaft. Specifically, the interior cylinder of the wheel plus the shaft became the *a*-partition and the rest of the wheel and the blades belonged to the *o*-partition, as explained in Chapter 4, Section 4.5. The blade motions are not considered.

FIGURE 8.7
Reduction component.

The remaining model size became 11,064 degrees of freedom. Of this, 10,764 degrees of freedom was the retained structural component (the shaft), and 300 mode shapes were representing the dynamic behavior of the reduced part of the structure.

The dynamic reduction method required the execution of the partitioning of the global matrices and the solution of an interior eigenvalue problem, with the latter being the more expensive. The real eigenvalue problem was also solved by the Lanczos method and the computational components required the following times:

Matrix factorization: 20.1 s

Linear system solution: 73.0 s

Orthogonalization: 0.48 s

Eigenvalue solution and eigenvector computation: 83.3 s

The total CPU time was 182.9 s, which is the sum of the previous items and some smaller unlisted activities. The elapsed time was 4 minutes and 39 seconds and this was only done once.

The CPU seconds of the reduced-order (11,064 degrees of freedom) complex eigensolution were

Matrix factorization: 7.4 s

Linear system solution: 21.5 s

Orthogonalization: 0.6 s

Eigenvalue solution: 1.2 s

Eigenvector computation: 6.0 s

The total CPU time of a complex eigenvalue computation was 36.3 s and the elapsed time was only 2 minutes and 28 seconds per RPM. Even though the analysis was executed in 100-RPM steps, and even with the consideration of the one-time cost of the reduction computation, the total time was only about 350 minutes, enabling the execution of such analyses overnight, thus fitting into the daily design cycle of engineering companies. In real-life turbine models, multiple stages such as the above occur; hence the time requirements are even more serious.

Of course one needs to consider the loss of accuracy occurring when using any reduced-order method. The numerical accuracy comparison of the two solutions is demonstrated in Figure 8.8 by comparing the eigenvalues found by the two methods at 2,500 RPM. The x represents the reduced-order solution, and the + represents the direct solution. As the figure shows, the complex eigenvalues, at least the first few dozen, are nearly undistinguishable, attesting to the value of the reduction method. Their accuracy somewhat diminishes later, a fact that may be rectified by increasing the number of mode shapes representing the dynamically reduced partition.

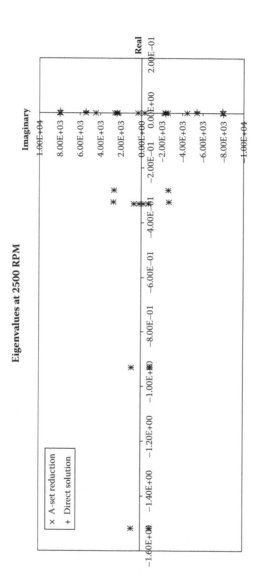

FIGURE 8.8
Accuracy comparison.

The reduced-order methods provide the engineer with the option of executing analyses at many more rotational speed steps, but care must be taken in selecting the reduced and retained parts of the structure. Inappropriate reduction may produce incorrect results and could miss important critical speed phenomena.

The reducible component also depends on the physical circumstances, especially those of the bearing. In the above case a simple spring-damper bearing was used that retains the symmetricity of the stiffness matrix. Hence the dynamic reduction of the wheel and blades was executed by a real eigenvalue analysis. Industrial speed-dependent bearings, such as the journal bearing described in the next section, result in unsymmetric stiffness matrices, in which case the complex dynamic reduction shown in Chapter 4, Section 4.6 is applicable.

8.4 The Journal Bearing

In practice, the attachment of the rotor to the stationary environment is via a bearing. In simple examples this is modeled by connecting points on the rotor and on the stationary part via spring and damper finite elements. The industrial bearings of rotating machinery, however, are more complex and their effects are encapsulated in the bearing-induced matrices (K_D, D_E) of the finite element equations.

Here we consider the so-called journal bearing shown in Figure 8.9, comprised of a cylindrical outer part and a rotating shaft with an oil film between them. The radius of the bearing and the shaft (journal) are denoted by R and r. The radial clearance is then $c = R - r$, the difference between the two radii.

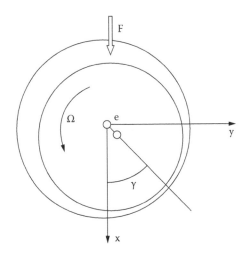

FIGURE 8.9
Journal bearing.

The shaft may be loaded by the force of the gravity or a pressure force statically (*F*), and it is pressed down to the bottom of the bearing in a stationary position. When rotation starts the journal is carried by the oil film and is displaced. The position of the bearing is defined by the eccentricity *e* and the attitude angle γ as shown in the figure.

These are calculated by solving the Reynolds equation describing the physical phenomenon at hand. In the following we will mainly rely on Gasch et al. [6], who gave an analytical solution to the Reynolds equation. Someya [7] solved the same problem numerically with different methods for several standard bearing types, and we will use this reference as validation for the analytic method.

An important parameter for journal bearings is the Sommerfeld number, which is a function of geometry, force, radial clearance, oil viscosity, and rotor speed. In ref. [6] the Sommerfeld number is defined in the "German" notation as

$$\bar{S} = \frac{F\psi^2}{LD\eta\Omega},$$
(8.1)

and in ref. [7] the definition is as follows:

$$S = \frac{\eta NDL}{\psi^2 F}.$$
(8.2)

The physical characteristics of the bearing are as follows: *L*, the width of bearing; *D* = 2*R*, the diameter; and η, the dynamic viscosity of the oil. The relative radial clearance is

$$\psi = \frac{c}{R},$$

and

$$\beta = \frac{L}{D}$$

is the relative width, while *F* is the aforementioned static force.

$$N = \frac{\Omega}{2\pi}$$

is the rotor speed in revolutions per second; hence the relationship between the two definitions is

$$S = \frac{1}{2\pi\bar{S}}.$$
(8.3)

Ref. [6] expresses the relationship between the Sommerfeld number and the relative eccentricity ε = *e*/*c* as

$$\bar{S}(\Omega) = \beta^2 \frac{\pi}{2} \frac{\varepsilon}{(1-\varepsilon^2)^2} \sqrt{(1-\varepsilon^2) + \left(\frac{4\varepsilon}{\pi}\right)^2}.$$
(8.4)

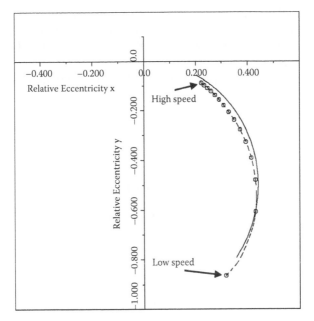

FIGURE 8.10
Position of the journal in the bearing.

This equation may be solved numerically for the eccentricity ratio ε as a function of the rotor speed Ω. The attitude angle is found in ref. [6] as

$$\tan \gamma = \frac{\pi}{4}\sqrt{\frac{1-\varepsilon^2}{\varepsilon^2}}. \tag{8.5}$$

With these two values the position of the journal in the bearing is known. The resulting position of the journal is shown in Figure 8.10 for different rotor speeds, where the increased frequency of the points refers to the higher RPM values.

It may be observed that at higher RPM values the journal is closer to the neutral position (0, 0) and at lower speeds the journal is closer to the bottom point (0, –1). The solid line is obtained from ref. [6], and the dashed lines are obtained from ref. [7].

The figure represents the results for a bearing with $D = 80$-mm diameter and $L = 40$-mm length with a clearance of 0.2 mm. The tables in ref. [7] were established for a relative width of 1 ($L = D$) and for a bearing with two axial groves for the oil flow in horizontal positions. The equations in ref. [6] are valid for a plain "short" bearing with $L < D$. The agreement between both methods is relatively good considering the example does not satisfy either of the conditions exactly.

The physical stiffness and damping values of the bearings are calculated from the eccentricity in terms of dimensionless quantities as

$$k_{ij} = \frac{F}{c}\gamma_{ij}, \, d_{ij} = \frac{F}{\Omega c}\beta_{ij}, \tag{8.6}$$

for all combinations of $i, j = x, y$. Here, k and d are the physical stiffness and damping values, and γ_{ij}, β_{ij} are the dimensionless values. Note that the stiffness is not dependent explicitly on the rotational speed, but implicitly through the relative eccentricity.

In ref. [6] the following expressions were found for the dimensionless stiffness and damping values:

$$\gamma_{xx} = Q(\varepsilon)\frac{\pi^2 + (32 + \pi^2)\varepsilon^2 + (32 - 2\pi^2)\varepsilon^4}{1 - \varepsilon^2}, \tag{8.7}$$

$$\gamma_{xy} = Q(\varepsilon)\frac{\pi}{4}\frac{\pi^2 + (32 + \pi^2)\varepsilon^2 + (32 - 2\pi^2)\varepsilon^4}{\varepsilon\sqrt{1 - \varepsilon^2}}, \tag{8.8}$$

$$\gamma_{yx} = -Q(\varepsilon)\frac{\pi}{4}\frac{\pi^2 - 2\pi^2\varepsilon^2 - (16 - \pi^2)\varepsilon^4}{\varepsilon\sqrt{1 - \varepsilon^2}}, \tag{8.9}$$

$$\gamma_{yy} = Q(\varepsilon)[2\pi^2 + (16 - \pi^2)\varepsilon^2]. \tag{8.10}$$

$$\beta_{xx} = Q(\varepsilon)\frac{\pi}{2}\frac{\pi^2 + (48 - 2\pi^2)\varepsilon^2 + \pi^2\varepsilon^4}{\varepsilon\sqrt{1 - \varepsilon^2}}, \tag{8.11}$$

$$\beta_{xy} = \beta_{yx} = Q(\varepsilon)\ [2\pi^2 + (4\pi^2 - 32)\varepsilon^2], \tag{8.12}$$

$$\beta_{yy} = Q(\varepsilon)\ \frac{\pi}{2}\sqrt{\frac{1 - \varepsilon^2}{\varepsilon^2}}\ [\pi^2 + (2\pi^2 - 16)\varepsilon^2], \tag{8.13}$$

where

$$Q(\varepsilon) = \frac{4}{[\pi^2 + (16 - \pi^2)\varepsilon^2]^{3/2}}. \tag{8.14}$$

The dimensionless stiffness values as a function of the relative eccentricity are shown in Figure 8.11, and the dimensionless damping values are shown in Figure 8.12. In both figures, the solid lines are from ref. [6] and the dashed lines with symbols are from ref. [7]. They are again in good agreement, apart from some minor differences in damping.

Let us calculate our bearing example introduced earlier. As mentioned, the diameter is 80 mm, the length in 40 mm, and the radial clearance is 0.2 mm. A force of 2452.5 N is applied and the oil viscosity is

$$\eta = 6.7588\text{E-8}\frac{\text{Ns}}{\text{mm}^2}.$$

Hence the physical stiffness and damping values can be calculated from Equation (8.6) using the dimensionless values. They are shown in Figures 8.13 and 8.14

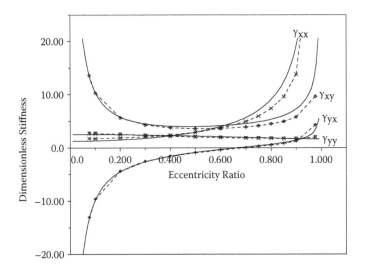

FIGURE 8.11
Dimensionless stiffness.

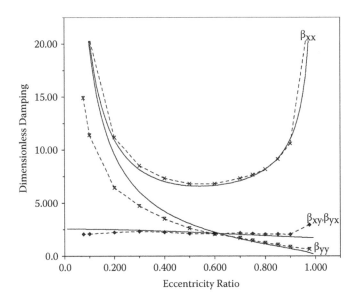

FIGURE 8.12
Dimensionless damping.

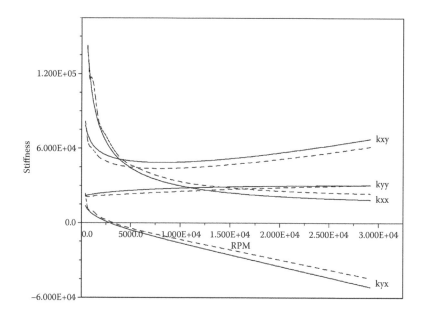

FIGURE 8.13
Bearing stiffness as a function of rotation speed.

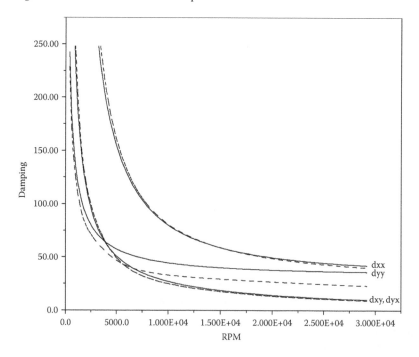

FIGURE 8.14
Bearing damping as a function of rotation speed.

as a function of the rotor speed. The solid lines are again from ref. [6] and the dashed lines are from ref. [7]. The agreement between the two methods is good, with some minor differences in the sideways damping.

The incorporation of this journal-bearing model into the finite element system is our final concern. The computed (speed-dependent) terms are going to be mapped to the appropriate degrees of freedom of the model as shown in Equation (8.15):

$$
K_D = \begin{bmatrix}
\begin{matrix} k_{xx} & k_{xy} \\ k_{yx} & k_{yy} \end{matrix} & & \begin{matrix} -k_{xx} & -k_{xy} \\ -k_{yx} & -k_{yy} \end{matrix} & \\
& 0 & & 0 \\
& & \ddots & \\
\begin{matrix} -k_{xx} & -k_{xy} \\ -k_{yx} & -k_{yy} \end{matrix} & & \begin{matrix} k_{xx} & k_{xy} \\ k_{yx} & k_{yy} \end{matrix} & \\
& 0 & & 0 \\
& & & & \ddots
\end{bmatrix}
$$

(8.15)

Here the location of the positive stiffness values is at the connecting node points of the bearing and the negative values represent their couping. The sparsity pattern shown in the equation is illustrative; the dots indicate the parts of the structure unrelated to the bearing connection.

The damping terms are mapped into the external damping matrix similarly and at the same locations, but here the matrix is symmetric with $d_{yx} = d_{xy}$:

$$
D_E = \begin{bmatrix}
\begin{matrix} d_{xx} & d_{xy} \\ d_{yx} & d_{yy} \end{matrix} & & \begin{matrix} -d_{xx} & -d_{xy} \\ -d_{yx} & -d_{yy} \end{matrix} & \\
& 0 & & 0 \\
& & \ddots & \\
\begin{matrix} -d_{xx} & -d_{xy} \\ -d_{yx} & -d_{yy} \end{matrix} & & \begin{matrix} d_{xx} & d_{xy} \\ d_{yx} & d_{yy} \end{matrix} & \\
& 0 & & 0 \\
& & & & \ddots
\end{bmatrix}
$$

(8.16)

The components containing zero might be used for a constant, uncoupled axial directional effect of the bearing (k_{zz}, d_{zz}). Such could occur in the case of water turbines, but its details are not followed here.

Let us now return to our single mass point and assume that it is supported by two bearings with translational motion only and dimensionless speed-dependent characteristics. The equation of motion for the particle is

$$
\left(\lambda^2 m \begin{bmatrix} 1 & \\ & 1 \end{bmatrix} + \lambda \frac{2F}{c\Omega} \begin{bmatrix} \beta_{xx} & \beta_{xy} \\ \beta_{yx} & \beta_{yy} \end{bmatrix} + \frac{2F}{c} \begin{bmatrix} \gamma_{xx} & \gamma_{xy} \\ \gamma_{yx} & \gamma_{yy} \end{bmatrix} \right) \begin{Bmatrix} u_x \\ u_y \end{Bmatrix} = \begin{Bmatrix} 0 \\ 0 \end{Bmatrix}.
$$

(8.17)

The coefficients are defined in terms of the force and the radial clearance. Dividing by the coefficient of the displacement term results in

$$
\left(\lambda^2 \frac{mc}{2F} \begin{bmatrix} 1 & \\ & 1 \end{bmatrix} + \lambda \frac{1}{\Omega} \begin{bmatrix} \beta_{xx} & \beta_{xy} \\ \beta_{yx} & \beta_{yy} \end{bmatrix} + \begin{bmatrix} \gamma_{xx} & \gamma_{xy} \\ \gamma_{yx} & \gamma_{yy} \end{bmatrix} \right) \begin{Bmatrix} u_x \\ u_y \end{Bmatrix} = \begin{Bmatrix} 0 \\ 0 \end{Bmatrix}.
$$

(8.18)

We introduce a reference frequency of the system as

$$
\omega_0^2 = \frac{2F}{mc},
$$

(8.19)

a dimensionless rotor speed

$$
v = \frac{\Omega}{\omega_0},
$$

(8.20)

and a dimensionless eigenvalue

$$
\bar{\lambda} = \frac{\lambda}{\omega_0}.
$$

(8.21)

Substituting these results in the quadratic eigenvalue problem of

$$
\left(\bar{\lambda}^2 \begin{bmatrix} 1 & \\ & 1 \end{bmatrix} + \bar{\lambda} \frac{1}{v} \begin{bmatrix} \beta_{xx} & \beta_{xy} \\ \beta_{yx} & \beta_{yy} \end{bmatrix} + \begin{bmatrix} \gamma_{xx} & \gamma_{xy} \\ \gamma_{yx} & \gamma_{yy} \end{bmatrix} \right) \begin{Bmatrix} u_x \\ u_y \end{Bmatrix} = \begin{Bmatrix} 0 \\ 0 \end{Bmatrix}.
$$

(8.22)

Note that the dimensionless stiffness and damping values in this equation are dependent on the eccentricity as was shown in Figures 8.11 and 8.12. The analytic solution of the above equation is shown in Figures 8.15 and 8.16 with the

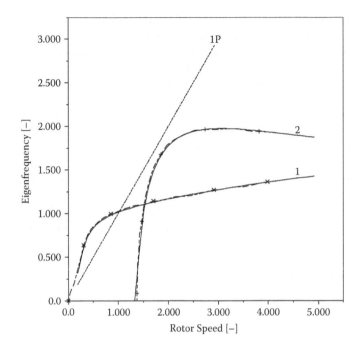

FIGURE 8.15
Campbell diagram.

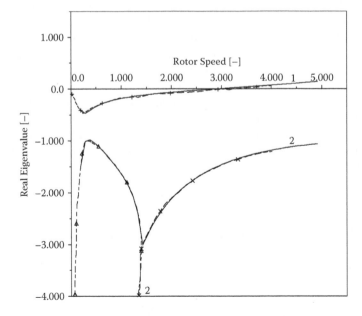

FIGURE 8.16
Real part.

solid curves. The first is the Campbell diagram and the second is the real part of the two dimensionless (demonstrated by the lack of unit in the notation [-]) eigenvalues. The dashed lines with symbols represent values from ref. [6] and they are in good agreement with the analytic solution.

The Campbell diagram in Figure 8.15 indicates that the model has a critical speed at 1.0 of the dimensionless eigenvalue, or at ω_0 of the absolute value. The real eigenvalue diagram in Figure 8.16 also indicates the beginning of an instability region at 2.98 of the dimensionless rotation speed that corresponds to a dimensionless eigenfrequency of 1.28.

The presence of both of these demonstrate the influence of the damping when comparing this with the result obtained in Chapter 6, Section 6.3, where the rotating mass point was only connected to two constant springs and no dampers. In that case we obtained two critical speed values, but no instability. Clearly, the instability arose from the damping introduced by the bearing.

8.5 Active External Loads

We conclude this chapter by considering the inclusion of active loads on the structure into the finite element equation as shown in Chapter 3, Section 3.3. This may be necessary when the rotating structure operates in an external field representing a physical phenomenon surrounding the structure, for example, when a turbine operates in a fluid flow or a wind turbine in an airflow. There are distinct differences between the two models (short, rigid vs. long, flexible blades) and we will discuss the latter in the next chapter in detail. Here we focus on the turbine in a fluid flow. We will assume here that the external flow is represented by a set of forces at certain spatial locations that are not necessarily coincident with the node points of the finite element model. These forces may be computed from the pressure values of the computational solution of the external fluid flow. That solution may be of a finite volume or finite difference technology.

Coupling those forces to the surface of the rotating structure is based on fitting a surface spline through a set of independent points given on the structural surface as

$$(x_i, y_i), i = 1, 2, \ldots n. \tag{8.23}$$

We restrict this discussion first to a surface component on a face perpendicular to the axis of rotation, that is, a plane parallel to the x-y plane. The surface spline is of the form

$$u(x, y) = a_0 + a_1 x + a_2 y + \sum_{i=1}^{n} K_i(x, y) P_i, \tag{8.24}$$

where P_i are virtual concentrated forces to achieve the desired deformation (u) of the surface to match a set of dependent points from the external principle. The K spring constants (see ref. [8] for more details) are

$$K_i(x,y) = \frac{1}{16\pi D} r_i^2 \ln r_i^2,$$

where D is the plate rigidity coefficient and

$$r_i^2 = (x - x_i)^2 + (y - y_i)^2.$$

The matrix form of the spline equation is

$$u = \begin{bmatrix} 0 \\ 0 \\ 0 \\ u_1 \\ \cdots \\ u_n \end{bmatrix} = \begin{bmatrix} 0 & 0 & 0 & 1 & \cdots & 1 \\ 0 & 0 & 0 & x_1 & \cdots & x_n \\ 0 & 0 & 0 & y_1 & \cdots & y_n \\ 1 & x_1 & y_1 & 0 & \cdots & K_{1n} \\ 1 & \cdots & \cdots & \cdots & 0 & \cdots \\ 1 & x_n & y_n & K_{n1} & \cdots & 0 \end{bmatrix} \begin{bmatrix} a_0 \\ a_1 \\ a_2 \\ P_1 \\ \cdots \\ P_n \end{bmatrix} = AC. \qquad (8.25)$$

In the above equation

$$K_{ik} = K_i(x_k, y_k); k = 1, 2, \dots, n.$$

Hence the solution for the unknown (boundary) coefficients (a) and forces (P) is

$$C = \begin{bmatrix} a_0 \\ a_1 \\ a_2 \\ P_1 \\ \cdots \\ P_n \end{bmatrix} = A^{-1} \begin{bmatrix} 0 \\ 0 \\ 0 \\ u_1 \\ \cdots \\ u_n \end{bmatrix} = A^{-1}u. \qquad (8.26)$$

The approximation at the given set of dependent (external force) points

$$(\bar{x}_j, \bar{y}_j), j = 1, 2, \dots m \qquad (8.27)$$

in terms of the just-computed coefficients and virtual forces becomes

$$\bar{u} = \begin{bmatrix} 1 & \bar{x}_1 & \bar{y} & K_{1,1} & & K_{1,n} \\ ... & ... & ... & ... & ... & ... \\ 1 & \bar{x}_m & \bar{y}_m & K_{m,1} & & K_{m,n} \end{bmatrix} \begin{bmatrix} a_0 \\ a_1 \\ a_2 \\ P_1 \\ ... \\ P_n \end{bmatrix} = BC. \tag{8.28}$$

In the above equation the first subscript of K is with respect to the j (dependent) index and the second subscript is with respect to the i (independent) index; hence the system of equations is rectangular. Substitution of the C matrix results in

$$\bar{u} = BA^{-1}u, \tag{8.29}$$

which establishes the transformation matrix enforcing the common surface spline fit as

$$G_{mn} = BA^{-1}. \tag{8.30}$$

Here the indices indicate the rectangular nature and the row and column sizes, respectively. The forces from the external (fluid pressure) field are transformed onto the structural locations by the transpose of this matrix as

$$F_n = G_{mn}^T F_m. \tag{8.31}$$

The transformation matrix could be computed directly in transposed form as follows:

$$G_{mn}^T = A^{-T}B^T. \tag{8.32}$$

The inverse matrix is not formed explicitly. The above operation is executed via a linear solve and the right-hand side contains only given or precomputed terms.

We now remove the restriction requiring that the surface must be parallel to the x-y plane, or strictly perpendicular to the axis of rotation. The blades of the turbine are curved, in fact, double curved, surfaces. Their curvature, however, is limited and there is a distinct side facing the flow.

Because the computer-aided surface modeling tools use parametric surfaces, we can extend the above methodology to the surface of these blades. Such parametric surfaces are the B-spline surfaces, and their generic form is

$$S(p,q) = \sum_{j=0}^{nj} \sum_{k=0}^{nk} B_j(p)B_k(q)Q_{jk}, \tag{8.33}$$

where the $B_{j,k}$ are basis functions of a certain degree, mostly cubic. Q_{jk} are so-called control points defining the shape of the surface. More on this may be found in ref. [9]. Any point on the curved surface, like the points of the blade surface facing the flow, may be described by the two parameters of the surface point:

$$(x_i, y_i, z_i) = S(p_i, q_i), i = 1, 2, ... n. \tag{8.34}$$

Then we select the fluid solution points in the neighborhood of the spatial coordinates of these points and project them to the parametric surface, producing

$$(\overline{x}_i, \overline{y}_i, \overline{z}_i) => (\tilde{x}_i, \tilde{y}_i, \tilde{z}_i), i = 1, 2, ... m. \tag{8.35}$$

Finally, the corresponding parametric coordinates of these points are found:

$$(\tilde{x}_i, \tilde{y}_i, \tilde{z}_i) = S(\overline{p}_i, \overline{q}_i), i = 1, 2, ... m. \tag{8.36}$$

The above coupling procedure is now simply executed in terms of these parametric coordinates with the $(x_i, y_i), (\overline{x}_i, \overline{y}_i)$ coordinate pairs replaced by the $(p_i, q_i), (\overline{p}_i, \overline{q}_i)$ coordinate pairs.
    We compute the A matrix of the transformation for a blade as

$$A = \begin{bmatrix} 0 & 0 & 0 & 1 & ... & 1 \\ 0 & 0 & 0 & p_1 & ... & p_n \\ 0 & 0 & 0 & q_1 & ... & q_n \\ 1 & p_1 & q_1 & 0 & ... & K_{1n} \\ 1 & ... & ... & ... & 0 & ... \\ 1 & p_n & q_n & K_{n1} & ... & 0 \end{bmatrix},$$

where the terms $K_{ik} = K_i(p_k, q_k); k = 1, 2, ..., n$ are also parametrically defined.
    The B matrix of the transformation may be computed as

$$B = \begin{bmatrix} 1 & \overline{p}_1 & \overline{q}_1 & K_{1,1} & & K_{1,n} \\ ... & ... & ... & ... & ... & ... \\ 1 & \overline{p}_m & \overline{q}_m & K_{m,1} & & K_{m,n} \end{bmatrix}.$$

Then the transformation matrix for the blade, $G_{mn}^T = A^{-T} B^T$, will convert the external forces to structural forces at the nodes.

9

Analysis of Aircraft Propellers

This chapter starts with a detailed description of a very generic model of a propeller blade. This is followed by the discussion of the airflow environment in which these machines are operating. Both quasi-steady and unsteady airflow are investigated. The final section discusses a four-bladed propeller application in an aircraft.

9.1 A Propeller Blade

Let us discuss the details of modeling and analyzing propeller blades, a topic of importance for wind turbines (the subject of the next chapter) and propeller-driven aircraft. Propeller blades are elastic and can be modeled by means of beam elements, but often shell or solid elements are also used. Due to their properties, the analysis of propellers must be done in the rotating system, and the geometric stiffness must be accounted for.

A typical propeller blade section is shown in Figure 9.1. The horizontal axis is the chord, and the vertical axis is the thickness of the blade at a certain chord location (c). The left tip of the profile is the leading edge (LE) and the right tip is the trailing edge (TE). The horizontal line connecting them contains several points of importance.

The quarter-chord point, the leftmost circle in the figure, is important for aeroelastic analysis. The second circle is the elastic axis (EA), which is obtained by connecting the shear centers of the radial cross-sections. When a force is acting vertically at this point, there will be no torsion. The third circle shows the location of the center of mass (CM), which is frequently located aft (toward the trailing edge) of the shear center. Finally, the fourth circle, denoted by R inside the profile in the figure, is the reference axis. This is usually the pitch axis of the blade.

The cross-section of the blade varies depending on its radial distance from the axis of rotation. The geometry of a propeller blade is typically described by plotting the locations of the leading edge (LE), trailing edge (TE), quarter chord (1/4), center of mass (CM), and elastic axis (EA) as functions of the blade radius. These quantities for an example propeller are shown in Figure 9.2. The sign of the quantities is with respect to the reference axis as shown in Figure 9.1. Clearly, the trailing edge is always aft of the reference

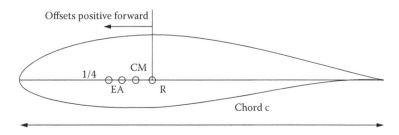

FIGURE 9.1
Typical cross section of a blade.

axis by definition; hence its sign is always negative, independent of the radial location of the section, as visible in Figure 9.2. The beginning of the blade is approximately at a radius of 25 cm. The outside edge of the blade is around 1.45 m.

The elastic stiffness distributions about the two profile axes and the torsion stiffness are shown in Figure 9.3 as the function of the radius. The interpretation is as intuition dictates: the blade is the stiffest in the proximity of the axis of rotation (at the hub). It is also very clear that the blade is significantly stiffer in the lag (in-plane) direction than in the flap (out of the plane of the rotation) direction.

The mass and mass moment of inertia of the cross-section, as functions of the radius, are shown in Figure 9.4. Another intuitive observation is that the mass and moment of inertia both decrease outward. Besides the radially changing profile, blades of propellers are also twisted in order to obtain an optimum aerodynamic angle of attack along the blades. This structural characteristic is the cause of the coupling between the flap and lag motions. An example blade twist is shown in Figure 9.5. The interpretation is that close

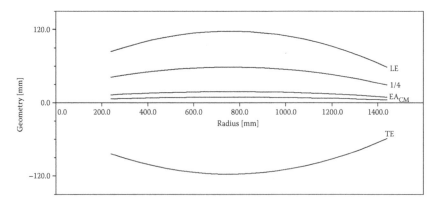

FIGURE 9.2
Blade geometry.

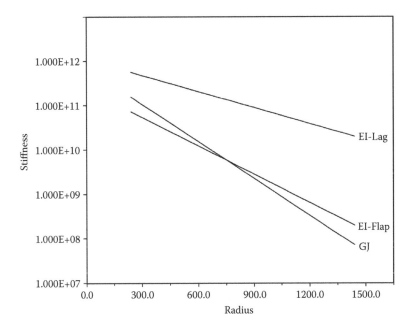

FIGURE 9.3
Stiffness distribution.

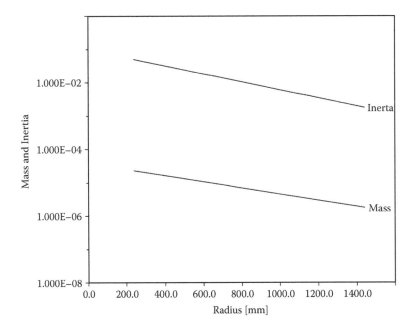

FIGURE 9.4
Mass and inertia distribution per length.

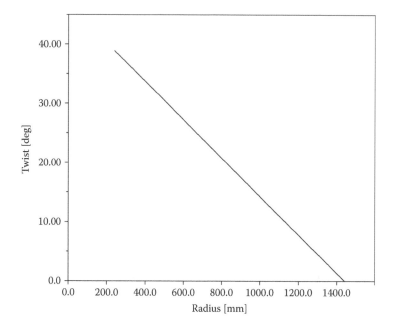

FIGURE 9.5
Blade twist.

to the hub the blade is twisted almost 40 degrees, and at the tip the blade cross-section is perpendicular to the axis of rotation.

Other important aspects to consider are the torsion of the blade, the relative positions of the center of mass, and the elastic axis. The latter are important in aeroelastic analysis, because the undesirable phenomenon of blade flutter can occur when the center of mass is aft of the elastic axis. An example twisted blade model is shown in Figure 9.6. The circles are shown to represent the mass distribution. The distance between the two points in the chord-wise direction is the radius of inertia. The midpoint between the circles is located at the line of the center of mass.

The Campbell diagram of the example blade is shown on Figure 9.7. The behavior of the modes is similar to those of the elastic arms discussed in Chapter 6, Section 6.8. The first flap mode (number 1) couples with the lag mode (number 2) around 3000 RPM. The coupling is manifested by the inflection of one of the participating curves. The first flap mode also intersects the 2P line on the Campbell diagram in Figure 9.7 at approximately 2000 RPM and the lag mode intersects at 1000 RPM.

The first three flap modes (numbers 1, 3, 4) of the blade are shown in Figures 9.8, 9.10, and 9.11. The first lag mode (number 2) or edgewise mode

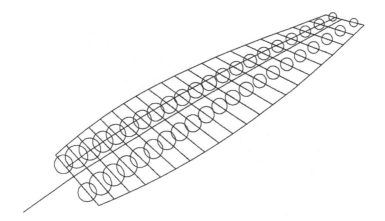

FIGURE 9.6
Finite element model of example blade.

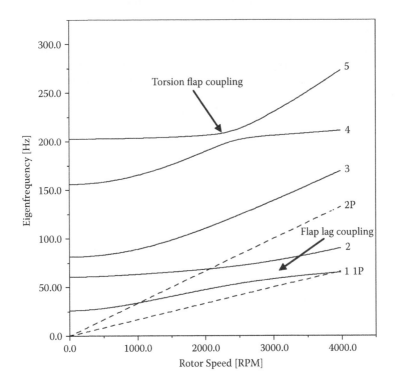

FIGURE 9.7
Campbell diagram of the propeller blade.

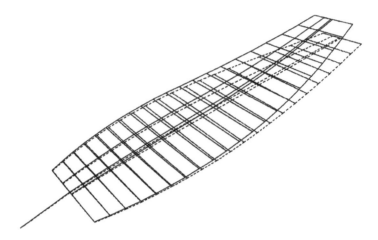

FIGURE 9.8
First flap mode at 26.1 Hz.

is shown in Figure 9.9. A coupling between the flap and lag motions can be seen in the plots. The torsion mode (number 5) is shown in Figure 9.12.

The Campbell diagram shows that the flap modes' frequencies are strongly increasing with the rotational speed. The torsion mode frequency is only slightly increasing with speed until it couples with the third flap mode around 2400 RPM. That could result in possible aerodynamic flutter instability.

The reason for the increase in frequency is the increase in stiffness, and this is explained in connection with Figure 9.13. The mass and inertia of the

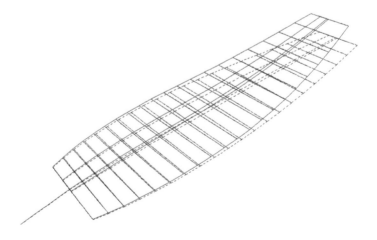

FIGURE 9.9
Lag mode coupled with flap at 60.8 Hz.

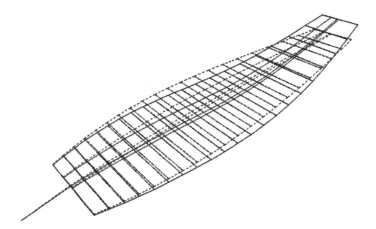

FIGURE 9.10
Second flap mode at 81.4 Hz.

cross-sections are represented by two mass points located at the radius of inertia, $i = \sqrt{\Theta/m}$. The centrifugal force acts radially outward from the axis of rotation and it has a value of $F = \Omega^2 i\, m/2$, as shown in the upper part of the figure looking along the rotor axis. When the blade is deformed in torsion by an angle φ, a pitching moment due to the y-component of the centrifugal force that acts against the deformation. It is equal to $M = \Omega^2 i\, m\varphi$ and occurs as shown in the lower part looking along the blade axis. The stiffness increase due to the propeller moment is $K = \Omega^2 \Theta$.

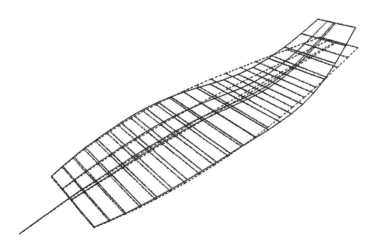

FIGURE 9.11
Third flap mode at 156.1 Hz.

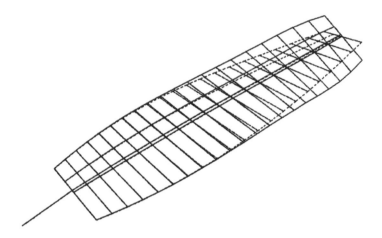

FIGURE 9.12
Blade torsion mode at 202.8 Hz.

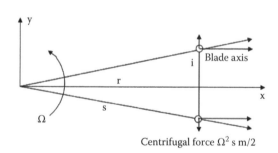

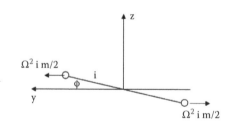

FIGURE 9.13
Propeller moment stiffness.

9.2 Quasi-steady Aerodynamics of Blade

During operation, rotating blades are not only affected by the rotor dynamic forces, but also by aerodynamic forces. We will investigate these forces in connection with a typical blade section at the radius r from the axis of rotation, as shown in Figure 9.14.

The tangential velocity arising from the rotation is Ωr, and the forward velocity of the air at the propeller is denoted by w. The angle between the inflow and the rotor plane is given by

$$\varphi = a \tan \frac{w}{\Omega r}. \tag{9.1}$$

The blade section is usually twisted and pitched in order to obtain the optimal angle of attack for different flight velocities. We denote the twist and pitch with the angle β, and the angle of attack is adjusted as

$$\alpha = \beta - \varphi. \tag{9.2}$$

The total air flow velocity becomes

$$V^2 = w^2 + \Omega^2 r^2. \tag{9.3}$$

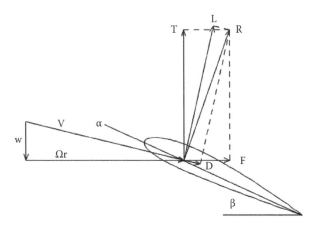

FIGURE 9.14
The velocity triangle of a propeller blade section.

The resulting aerodynamic lift, drag, and moment for a strip of the blade in the airflow, with width s and chord length c, are defined as

$$L = \frac{\rho}{2} V^2 cs C_L, \tag{9.4}$$

$$D = \frac{\rho}{2} V^2 cs C_D, \tag{9.5}$$

$$M = \frac{\rho}{2} V^2 c^2 s C_M. \tag{9.6}$$

The flow density is ρ. As shown in Figure 9.14, the lift force (L) is perpendicular to the flow direction and the drag acts in the flow direction. In the case of an aircraft propeller, the thrust T is now the contribution of the strip of the propeller blade to pull the aircraft. The engine must overcome the torque moment $M_r = Fr$. The power needed is then $P = \Omega M_r = \Omega Fr$. The total thrust and power can be found by integrating the strips over the length of the blade and summing up the blades. In the case of a wind turbine, the subject of the next chapter, the thrust T is the result of the wind flow and the energy generated by it is proportional to P.

The lift, drag, and moment coefficients depend on the profile used. An example of the coefficients as functions of the angle of attack is shown in Figure 9.15. The coefficients are given for the quarter-chord point location

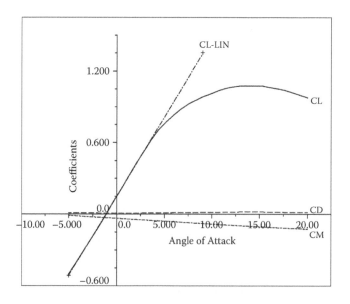

FIGURE 9.15
Aerodynamic coefficients for the quarter-chord point.

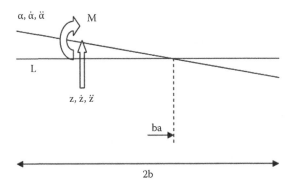

FIGURE 9.16
Unsteady lift and moment acting on a blade strip.

of the profile. In the normal operating range of small angles of attack, the coefficients can be linearized as follows:

$$C_L = C_{L0} + C_L'\alpha,$$

$$C_D = C_{D0},$$

$$C_M = C_{M0} + C_M'\alpha.$$

The influence of the drag is small and is neglected for the unsteady forces. The lift curve slope for a flat plate is $C_L' = 2\pi$. The linearized values are shown in Figure 9.15 as dash-dotted lines.

The lift and moment due to the motion of the blade section are shown in Figure 9.16. The degrees of freedom are the displacement normal to the profile z and the torsion about the elastic axis α. The chord length is $c = 2b$, and the elastic axis is located at the distance ba aft of the midpoint.

When calculating the unsteady forces, the static components are not considered. Hence the unsteady lift component due to the angle of attack is obtained as

$$L_\alpha = \rho V^2 s b C_L'\alpha. \tag{9.7}$$

The aerodynamic moment about the elastic axis to address to the unsteady lift force acts at a distance of $b\left(\frac{1}{2}+a\right)$ beyond the quarter-chord point, leading to

$$M_\alpha = \rho V^2 s b^2 \left(\tfrac{1}{2}+a\right)C_L'\alpha + 2\rho V^2 s b^2 C_M'\alpha. \tag{9.8}$$

In the last equation the reference chord of $2b$ is applied to adhere to the usual definition of the moment coefficient with the chord length as reference. The equations above represent the aerodynamic stiffness term.

The damping due to vertical velocity (the so-called plunge velocity) is derived from Figure 9.17. When the blade is moving up, the profile will

FIGURE 9.17
Aerodynamic damping.

encounter a downward velocity component $\dot{z}$. The angle of attack, there-fore, is $\alpha = \dot{z}/V$, and the velocity-dependent lift component becomes

$$L_{\dot{z}} = \rho V s b C'_L \dot{z}. \tag{9.9}$$

The moment with the lift and moment coefficients used is

$$M_{\dot{z}} = \rho V s b^2 (\tfrac{1}{2} + a) C'_L \dot{z} + 2\rho V s b^2 C'_M \dot{z}. \tag{9.10}$$

There is also a term arising from the downward velocity at the three-quarters point due to torsional velocity, as shown in Figure 9.18. This leads to the tor-sional damping terms for lift and moment:

$$L = C'_L \rho V s b^2 (\tfrac{1}{2} - a) \dot{\alpha}, \tag{9.11}$$

$$M = C'_L \rho V s b^3 (a + \tfrac{1}{2})(\tfrac{1}{2} - a) \dot{\alpha}. \tag{9.12}$$

Finally, all lift forces and moments of the strip can be combined in the matrix equation

$$\begin{Bmatrix} L \\ M \end{Bmatrix} = \rho V^2 b s C'_L \begin{bmatrix} 0 & 1 \\ 0 & b(\tfrac{1}{2}+a) \end{bmatrix} \begin{Bmatrix} z \\ \alpha \end{Bmatrix} + 2\rho V^2 b^2 s C'_M \begin{bmatrix} 0 & 0 \\ 0 & 1 \end{bmatrix} \begin{Bmatrix} z \\ \alpha \end{Bmatrix}$$

$$+ \rho V b s C'_L \begin{bmatrix} -1 & b(\tfrac{1}{2}-a) \\ -b(\tfrac{1}{2}+a) & b^2 (\tfrac{1}{2}-a)(\tfrac{1}{2}+a) \end{bmatrix} \begin{Bmatrix} \dot{z} \\ \dot{\alpha} \end{Bmatrix} + 2\rho V b^2 s C'_M \begin{bmatrix} 0 & 0 \\ 1 & 0 \end{bmatrix} \begin{Bmatrix} \dot{z} \\ \dot{\alpha} \end{Bmatrix}. \tag{9.13}$$

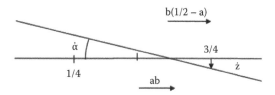

FIGURE 9.18
Velocity at the three-quarters point due to angular velocity about the elastic axis.

The first term is the aerodynamic stiffness due to lift. The matrix is unsymmetric because the force in the z-direction is caused by blade torsion, but there is no moment due to blade displacement. The third term is the aerodynamic damping also due to lift. This matrix is also unsymmetric. These terms are called circulatory terms because they are dependent on the lift caused by a circulation around the profile. The theory for unsteady aerodynamic forces can be found in aeroelastic textbooks such as those in, for example, refs. [10] through [14].

In Equation 9.13 the time lag between motion and force is not accounted for and the aerodynamic forces are called quasi-steady forces. That is not true in practice; a time delay is present due to the fact that the circulation adherent to a change in the angle of attack position needs some time to build up. That will be considered in the next section.

In addition to the circulatory forces, noncirculatory forces occur as well. When the profile is accelerated, the air around the profile is also accelerated. The apparent mass of air influenced in this fashion is equivalent to a cylinder of air with a diameter equal to the chord length, and the moment can be described by two cylinders with the half chord as diameters, as shown in Figure 9.19.

The inertia terms are important for extreme light structures. The apparent mass was recently verified during the ground vibration tests of the Solar Impulse aircraft and is published in Boeswald et al. [15]. Following this metric, the aerodynamic mass can be described by the following matrix equation:

$$\left\{ \begin{array}{c} L \\ M \end{array} \right\} = -\pi\rho b^2 s \left[\begin{array}{cc} 1 & ba \\ ba & b^2(a^2 + \frac{1}{8}) \end{array} \right] \left\{ \begin{array}{c} \ddot{z} \\ \ddot{\alpha} \end{array} \right\}. \tag{9.14}$$

There is also a noncirculatory term due to a centrifugal force caused by the torsion velocity:

$$\left\{ \begin{array}{c} L \\ M \end{array} \right\} = \pi\rho b^2 s\, V \left[\begin{array}{cc} 0 & 1 \\ 0 & b(\frac{1}{2} - a) \end{array} \right] \left\{ \begin{array}{c} \dot{z} \\ \dot{\alpha} \end{array} \right\}. \tag{9.15}$$

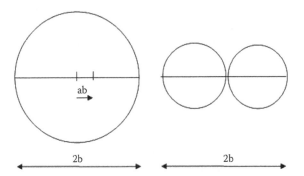

FIGURE 9.19
Apparent mass and inertia.

In these equations V is the resultant airflow velocity that we will define now. Using the tip radius R of the blade (the distance of the outermost point and the axis of rotation), the so-called advance ratio is defined as

$$\mu = \frac{w}{\Omega R}.$$

(9.16)

Introducing the dimensionless radius as

$$\eta = \frac{r}{R},$$

(9.17)

the resultant velocity is found to be

$$V^2 = \mu^2 \Omega^2 R^2 + \Omega^2 \eta^2 R^2 = \Omega^2 R^2 (\mu^2 + \eta^2).$$

(9.18)

Selecting a value for the advance ratio, the resultant velocity can be calculated for each strip and summed up for the whole blade. The displacement and torsion angle of the blade segment can be found by a transformation with the twist angle. Finally, the lift and moment of the blade element must be transformed to the rotor system.

The above forces can be added to the rotor dynamic equilibrium equations as aerodynamic mass, stiffness, and damping forces. In the case of quasi-steady forces the eigenvalue solution is straightforward because the aerodynamic matrices can simply be added to the structural mass, damping, and stiffness matrices. The result of a quasi-steady analysis is shown in Figure 9.20 for an advance ratio of 0.1. The results without aerodynamic forces were shown earlier in Figure 9.7 and are also shown as dashed lines. The corresponding result lines with aerodynamic forces are denoted by the symbol A. The analysis can be repeated for other advance ratios and pitch angles in order to study other operating conditions.

The aerodynamic stiffness is negative for blade torsion because the lift is acting in front of the elastic axis. Hence a positive torsion angle leads to a positive lift and a positive moment in the same sense as the displacement. Therefore the eigenfrequency of the torsion mode (5A) is decreasing with speed. Around 1800 RPM there is a weak coupling with the third flap modes (4A), and around 3000 RPM there is a coupling with the second flap mode (3A). There is also a coupling with the first flap (1A) and lead-lag modes (2A), and around 4000 RPM the lead-lag frequency goes to zero. Above this speed there are two real roots instead of one pair of complex conjugate roots. This is called divergence and can also occur for (nonrotating, fixed) aircraft wings when the negative aerodynamic stiffness gets larger than the structural stiffness. There it is called aeroelastic divergence.

The damping curves are shown in Figure 9.21, where we show only the aerodynamic force results; hence the A notation is discontinued. The torsion mode (5) gets unstable around 2100 RPM. This is called flutter instability,

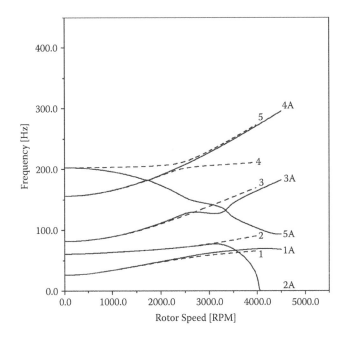

FIGURE 9.20
Eigenfrequencies of the blade with quasi-steady aerodynamic forces.

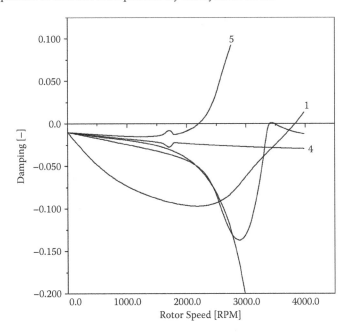

FIGURE 9.21
Damping values of the propeller blade with aerodynamic forces.

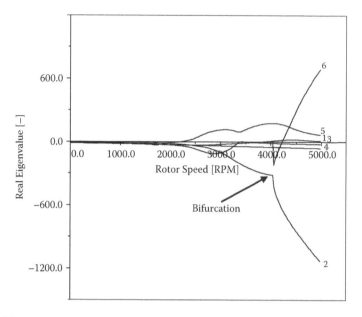

FIGURE 9.22
Real eigenvalues.

and it is an especially dangerous problem for fixed-wing aircraft. There is a coupling between the first flap (1) and lead-lag modes (2), and the flap mode (1) gets unstable around 3800 RPM. There is also a coupling between torsion (5) and the second flap mode (3), and the latter gets almost unstable around 3400 RPM for a small interval. The damping of the flap modes is, otherwise, rather high.

The real parts of the eigenvalues are shown in Figure 9.22. After the bifurcation point around 4000 RPM, there is a stable branch with increasing negative values denoted by 2. The branch denoted by 6 becomes unstable around 4200 RPM. This is the divergence instability.

The radius of the example propeller is 1.44 m, and, assuming a tip speed of 300 m/s, the operating rotor speed would be around 1900 RPM. The margin to the flutter speed of 2100 RPM is therefore very low. On the other hand, the divergence instability is far above the operating speed.

9.3 Unsteady Aerodynamics of Blade

We now consider the effect of the time lag between motion and force defined by the Wagner function shown in Figure 9.23. The Wagner function represents the delay of the lift due to a step function in the angle of attack. It is 0.5

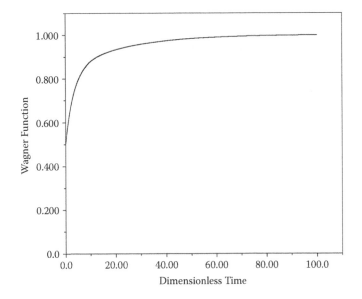

FIGURE 9.23
Wagner function defining the delay of the circulation due to a step.

when the time is zero and increases asymptotically to 1. The argument of the Wagner function is the dimensionless time quantity of the form

$$\tau = \frac{Vt}{b}.$$

The Wagner function can be approximated by

$$\Phi(\tau) = 1 - 0.165e^{-0.0455\tau} - 0.335e^{0.3\tau}.$$

The lift force coefficient in the time domain will be adjusted by this function as

$$C'_L(t) = 2\pi\Phi(t). \tag{9.19}$$

Let us now move from the time domain to the frequency domain, which is more amenable to our rotor dynamic focus. Executing a Laplace transform on the Wagner function results in the Theodorsen function shown in Figure 9.24. This function was derived by Theodorsen [16] and by Küssner and Schwarz [17] and is described in many textbooks on aeroelasticity. See, for example, Bisplinghoff, Ashley, and Halfman [10] and Rodden

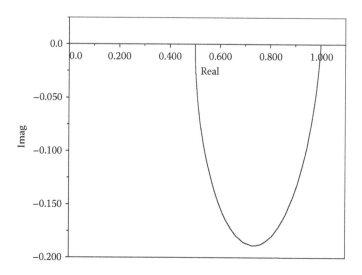

FIGURE 9.24
Theodorsen function.

[11]. The argument of the Theodorsen function is the reduced frequency defined by

$$k = \frac{\omega b}{V}.$$ (9.20)

The Theodorsen function is complex:

$$C(k) = F(k) + i\,G(k),$$ (9.21)

where the real part of the function is 1.0 at $k = 0$ and 0.5 when $k = \infty$. The real and imaginary components are defined as

$$F(k) = \frac{J_1^2(k) + J_1(k)Y_0(k) - J_0(k)Y_1(k) + Y_1^2(k)}{D(k)},$$ (9.22)

$$G(k) = -\frac{J_0(k)J_1(k) + Y_0(k)Y_1(k)}{D(k)},$$ (9.23)

and

$$D(k) = \left(J_1(k) + Y_0(k)\right)^2 + \left(J_0(k) - Y_1(k)\right)^2.$$ (9.24)

The cylindrical functions are the Bessel functions of first (J) and second (Y) kind with the subscript indicating order 0 or 1. The Theodorsen function is used to adjust the lift coefficient by $C(k)$, similar to the adjustment by the Wagner function.

With this, the unsteady force-based circulatory terms become

$$
\begin{Bmatrix} L \\ M \end{Bmatrix} = \rho V^2 b s C_L' C(k) \begin{bmatrix} 0 & 1 \\ 0 & b\left(\frac{1}{2}+a\right) \end{bmatrix} \begin{Bmatrix} z \\ \alpha \end{Bmatrix}
$$

$$
+ 2\rho V^2 b^2 s C_M' C(k) \begin{bmatrix} 0 & 0 \\ 0 & 1 \end{bmatrix} \begin{Bmatrix} z \\ \alpha \end{Bmatrix}
$$

$$
+ \rho V b s C_L' C(k) \begin{bmatrix} -1 & b\left(\frac{1}{2}-a\right) \\ -b\left(\frac{1}{2}+a\right) & b^2\left(\frac{1}{2}-a\right)\left(\frac{1}{2}+a\right) \end{bmatrix} \begin{Bmatrix} \dot{z} \\ \dot{\alpha} \end{Bmatrix}
$$

$$
+ 2\rho V b^2 s C_M' C(k) \begin{bmatrix} 0 & 0 \\ 1 & 0 \end{bmatrix} \begin{Bmatrix} \dot{z} \\ \dot{\alpha} \end{Bmatrix}. \tag{9.25}
$$

Because the $C(k)$ term is a function of k and in turn ω, the eigenvalue problem is now nonlinear in the reduced frequency. The solution ω is contained in the system matrices for both stiffness and damping. Therefore the eigenvalue problem must be solved by iteration.

The unsteady forces may be formulated as real and imaginary matrices. The real part is the sum of the stiffness and mass matrix, including the non-circulatory terms, which is

$$
[Q_R(k)]
$$

$$
= \rho V^2 s \left(b C_L' C(k) \begin{bmatrix} 0 & 1 \\ 0 & b\left(\frac{1}{2}+a\right) \end{bmatrix} + 2b^2 C_M' C(k) \begin{bmatrix} 0 & 0 \\ 0 & 1 \end{bmatrix} - \pi k^2 \begin{bmatrix} 1 & ba \\ ba & b^2\left(a^2+\frac{1}{8}\right) \end{bmatrix} \right). \tag{9.26}
$$

The imaginary part is the damping matrix

$$
[Q_I(k)] = \rho V^2 s k
$$

$$
\left(C_L' C(k) \begin{bmatrix} -1 & b\left(\frac{1}{2}-a\right) \\ -b\left(\frac{1}{2}+a\right) & b^2\left(\frac{1}{2}-a\right)\left(\frac{1}{2}+a\right) \end{bmatrix} + \frac{2}{V} b^2 C_M' C(k) \begin{bmatrix} 0 & 0 \\ 1 & 0 \end{bmatrix} + \pi \begin{bmatrix} 0 & 1 \\ 0 & b^2\left(\frac{1}{2}-a\right) \end{bmatrix} \right). \tag{9.27}
$$

These matrices are computed for each aerodynamic blade element. Note that the chord length is different for the different elements, and even the velocity is different due to their different distances from the axis of rotation. Hence a reference reduced frequency for the whole blade is first defined and then corrected for all elements by their actual chord length and velocity. In practice the reference radius of 0.75 R is commonly used.

Thus far we have assumed incompressible flow. For compressible flow further corrections to the lift curve slope are needed. The Prandl–Glauert correction to a blade element is

$$C'_L = C'_{L0} \frac{1}{\sqrt{1 - Ma_r^2}}, \tag{9.28}$$

where Ma_r is the local Mach number (velocity divided by the velocity of sound) at radius r.

The unsteady aerodynamic matrices will be dependent on the global Mach number Ma, as

$$C'_L = C'_{L0} \sqrt{1 - Ma^2 \left(1 - \frac{\eta^2}{\mu^2}\right)},$$

where $U^2 = \Omega^2 R^2 (\mu^2 + \eta^2)$ and μ, η are the advance ratio and dimensionless radius defined earlier.

The complex eigenvalue problem providing the Campbell diagram solution can now be written as

$$\left[\lambda^2 [M] + \lambda \left([D] + 2\Omega[C] - \frac{\rho \bar{c} V}{4k} [Q_I(k, Ma)] \right) \right.$$
$$\left. + \left([K] - \Omega^2 [Z] + \Omega^2 [K_G] + \Omega[K_D] - \frac{\rho V^2}{2} [Q_R(k, Ma)] \right) \right] \{\varphi\} = \{0\}. \tag{9.29}$$

This eigenvalue problem can only be solved by iteration.

Choosing a suitable reduced frequency (k) calculated from the real eigenvalue or from the solution at the previous rotor speed, the aerodynamic forces are calculated and the eigenvalue problem is solved. The resulting eigenfrequency ω is used to calculate a new value of the reduced frequency. Then the aerodynamic forces are recalculated and the eigenvalue problem is solved again. This procedure is repeated until the reduced frequency does not change, meaning that the solution is converged. This must be repeated for all eigenfrequencies. The process is also known as frequency matching.

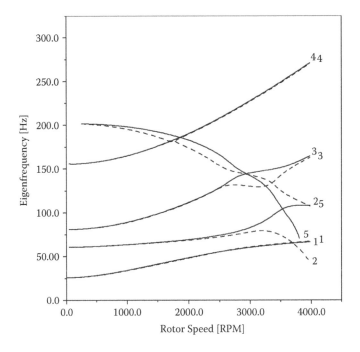

FIGURE 9.25
Eigenfrequencies with unsteady and quasi-steady aerodynamic forces.

The solution process is done for the selected rotor speeds and the results are postprocessed in the usual way to obtain frequency and damping plots. Usually, the aerodynamic matrices are calculated for a suitable range of reduced frequencies and Mach numbers. Then the actual matrices are calculated by interpolation in reduced frequency and Mach number. This solution method is known as the PK method and was developed by Hassig [18].

The eigenfrequencies from the PK iteration with unsteady aerodynamic forces are shown in Figure 9.25 as solid lines together with the results, with quasi-steady forces as dashed lines. The coupling between the torsion mode (5) and the second flap mode (3) is different between the quasi-steady and unsteady force case. The damping curves are shown in Figure 9.26. The critical flutter point has shifted up from 2100 to 2800 RPM. Now mode number 3 (solid curve) gets unstable instead of mode number 5 (dashed curve) for the quasi-steady solution. The flutter mechanism is, however, still a coupling between flap (3) and torsion (5), as shown in Figure 9.27. In the unsteady case there is a strong second instability around 3600 RPM for mode 2, the first lead-lag mode, which couples with flap and torsion.

The results of the PK method are shown in Figure 9.28. The constant reduced frequencies (k) appear as straight lines from the origin in the frequency diagram. The solution frequencies are plotted on top of that. Following a particular

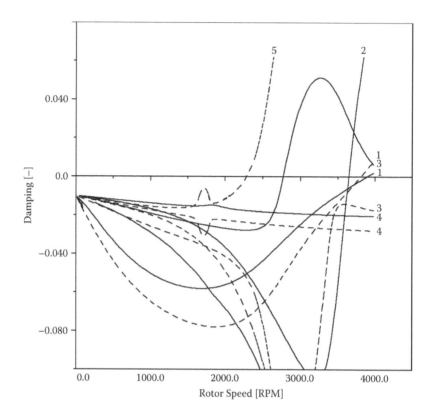

FIGURE 9.26
Damping curves with unsteady and quasi-steady aerodynamic forces.

eigensolution, for example, mode 5, we can see the gradual progress of the iteration. There is a good distribution in the whole area of rotor speed and frequency, which indicates a reliable solution. In the PK method the aerodynamic forces are calculated for several reduced frequencies. During the iteration an interpolation between the calculated matrices is done. Therefore a good distribution of the reduced frequencies in the solution space is necessary for convergence.

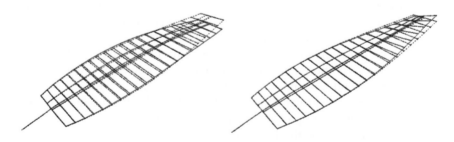

FIGURE 9.27
Real and imaginary parts of complex solution number 3 at 3000 RPM at 144 Hz.

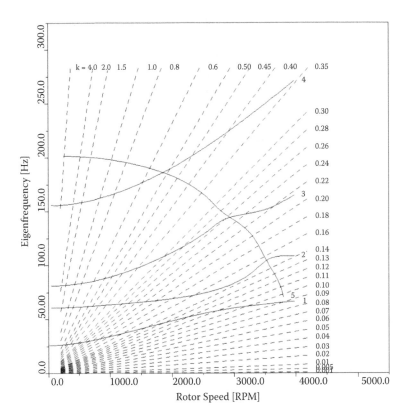

FIGURE 9.28
Solution of the PK method with reduced frequency lines.

The solution of the aeroelastic problem of a rotating blade, including unsteady aerodynamic forces, will also be applied for wind turbines in the next chapter.

9.4 Propeller with Four Blades

We will now extend the four-armed simple example from Chapter 6, Section 6.8 by replacing the arms with the blades shown in Figure 9.29. The blades have been pitched by 25 degrees in order to simulate an operating condition of 100-m/s flight speed and 1800-RPM rotor speed.

The Campbell diagram of this model is shown in Figure 9.30. We considered the shaft as being elastic, and the blade modes couple with the shaft as in the model with four arms. Furthermore, due to the pitching of the blades

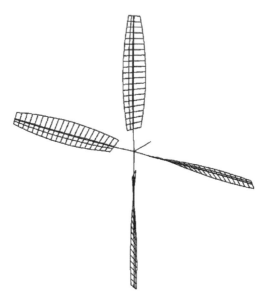

FIGURE 9.29
Propeller with four blades.

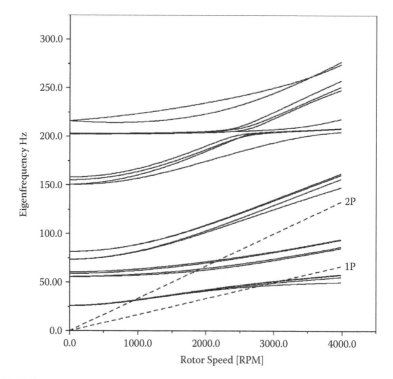

FIGURE 9.30
Campbell diagram of the propeller.

the blade modes are also turned and the resonance with the 1P line occurs around 3000 RPM.

Next we analyze the propeller shown in Figure 9.29 with quasi-steady aerodynamic forces. The eigenfrequencies are shown in Figure 9.31 and the damping is shown in Figure 9.32. It is noticeable that the critical speed is reduced compared to that shown in Figure 9.21 for the single blade. When such a propeller is attached to an aircraft, as shown in Figure 9.33, the velocity w of the prior sections is higher than the flight velocity because the propeller is accelerating the air stream. The additional velocity is called the induced velocity. This velocity increase can be found by taking the equilibrium between the thrust found from the blade elements and by the momentum theory of the propeller disk.

The induction can be described by the factor a as correction to the flight velocity w_0:

$$w = w_0(1+a). \tag{9.30}$$

The classical theory is described by H. Glauert [19] and the interested reader may follow this topic there.

Propellers attached to an aircraft must be calculated in the rotating system, but the aircraft structure must be calculated in the fixed system. Therefore the full coupled equation developed in Chapter 2, Section 2.4 must be used. This method is described for wind turbines in Chapter 10.

The aircraft model shown in Figure 9.33 is based on a half-model described in refs. [20] and [21] and was considerably modified for aeroelastic analyses in ref. [22].

Because the lowest eigenfrequency of a propeller blade is usually much higher than the wing and nacelle eigenfrequencies, the propeller can be regarded as rigid and attached to the aircraft, which can then be analyzed in the fixed system. Now we have two simple rotors like the one described in Chapter 6, Section 6.4 attached to a relatively soft structure. We assume a polar moment of inertia of 150 kgm^2 including engine effect. The turbine engine has a low moment of inertia but is running much faster, so the inertia is multiplied by the gear ratio and added to that of the propeller. Two typical modes without propeller rotations are shown in Figures 9.34 and 9.35. Both modes involve symmetric engine motion in pitch (vertical displacement) and yaw (sidewise displacement). Here we expect a gyroscopic coupling of the two modes.

The eigenfrequencies of the aircraft structure with rotating propellers are shown in Figure 9.36. Here we can see the influence of the propeller rotation. The eigenfrequency of mode number 8 is decreasing and that of mode number 10 is increasing with propeller speed. Modes 7 and 9 are antisymmetric modes and show a similar behavior. A similar effect was found for the Laval rotor in Chapter 6, Figure 6.7. Here the propeller inertia is small compared to

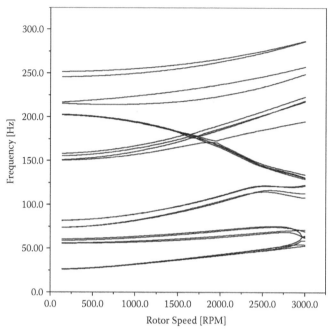

FIGURE 9.31
Campbell diagram of the propeller quasi-steady forces.

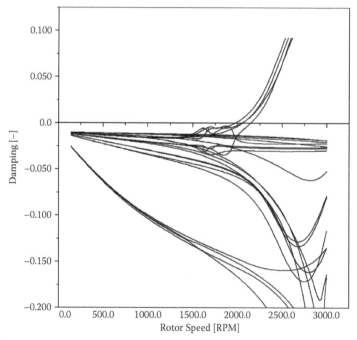

FIGURE 9.32
Damping curves for the propeller with quasi-steady forces.

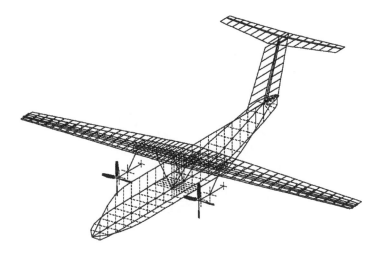

FIGURE 9.33
Transport aircraft with propellers.

the mass of the aircraft, and the effect of the propeller rotation on the aircraft modes is less pronounced.

Assuming that the propellers are rotating at 1800 RPM in the same sense of rotation, the complex modes are calculated and shown in Figures 9.37 and 9.38 for the real and the imaginary parts, respectively. Because the gyroscopic matrix is antisymmetric, there is a coupling between pitch and yaw motion, and the imaginary part of the mode is unsymmetric.

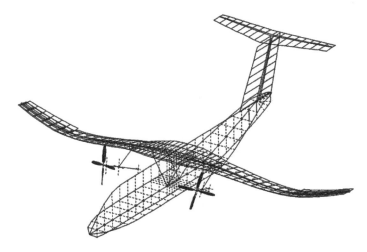

FIGURE 9.34
Mode number 8, symmetric wing torsion and engine pitch at 5.1 Hz.

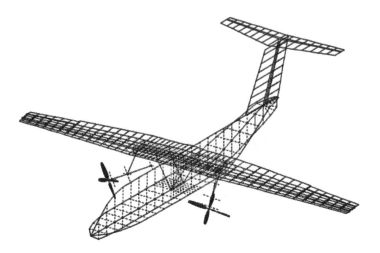

FIGURE 9.35
Mode number 10, symmetric engine yaw at 5.7 Hz.

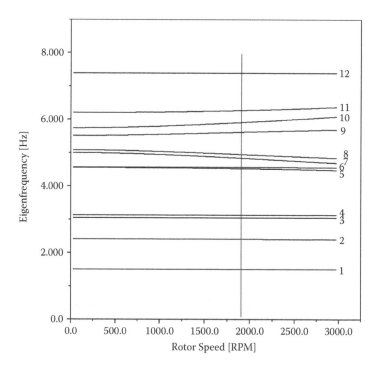

FIGURE 9.36
Eigenfrequencies of the aircraft structure with rotating propellers.

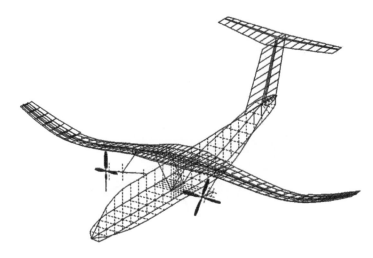

FIGURE 9.37
Real part of complex mode number 8. Propellers rotating in the same direction.

By changing the sense of rotation on the left engine so that the propellers are rotating in opposite directions (counter rotating) the real part is unchanged, as shown in Figure 9.39, but now the imaginary part, shown in Figure 9.40, is also symmetric.

In addition to the gyroscopic matrices, the aerodynamic whirl matrices must also be considered. After two severe accidents with the Lockheed Electra in 1959 and 1960 caused by engine/propeller whirl flutter, the

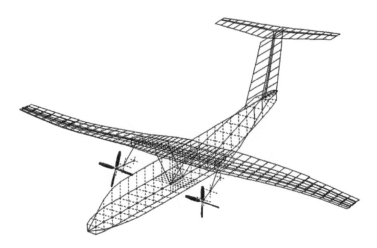

FIGURE 9.38
Imaginary part of complex mode number 8. Propellers rotating in the same direction.

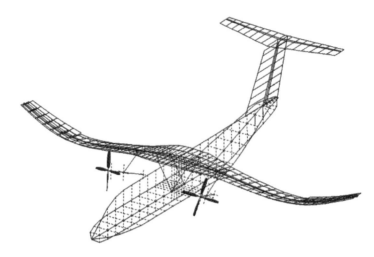

FIGURE 9.39
Real part of complex mode number 8. Propellers rotating in opposite directions.

problem was investigated and theories were developed for the calculation of
the aerodynamic forces of propellers with oblique inflow. See, for example,
refs. [23] through [27]. A description is also included in ref. [22]. In the case of
our example propeller on the example aircraft with aerodynamic propeller
terms included, the whirl instability of mode number 8 occurred at around

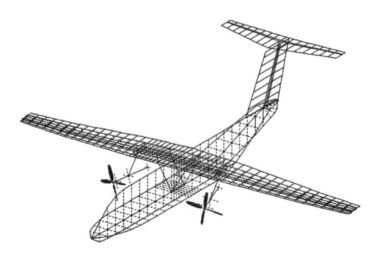

FIGURE 9.40
Imaginary part of complex mode number 8. Propellers rotating in opposite directions.

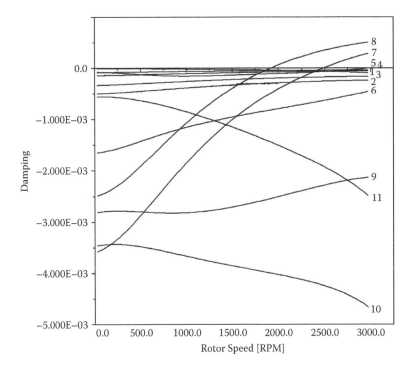

FIGURE 9.41
Damping diagram with aerodynamic propeller forces.

1800 RPM, as shown in Figure 9.41. A redesign of the engine pylons would be necessary to avoid this.

In addition to the gyroscopic terms and the aerodynamic forces of the propellers, the unsteady aerodynamic forces of the wings must also be considered in a whirl flutter analysis. This topic, however, is also beyond the scope of this book.

10

Analysis of Wind Turbines

Wind has become an important source of energy during the last 20 years. By 2010 there was about 200 GW of wind turbine capacity installed worldwide and the energy production reached 430 TWh. This is around 2.5% of the worldwide consumption of electric energy, and it is equivalent to more than 50 blocks of nuclear reactors. The largest turbines in use today have a 7-MW rated energy output and rotor diameters up to 125 m, both on land and in offshore installations. A description of the wind energy technology can be found in ref. [28].

There are even larger turbines in development with 70-m long blades, mainly for future offshore installations. Turbines of this size pose a great challenge for industrial analysis. Because the unsteady aerodynamic forces must also be included in the analysis, the coupled rotor dynamic and aeroelastic problem must be solved.

Industrial wind turbines may contain two blades, but most installations use three blades. Most wind turbines have a horizontal axis of rotation, but some vertical axis installations also occur in urban environments. Hence we will focus on horizontal axis, three-bladed turbines in the bulk of this chapter with a brief visit to the two-bladed structures at the end.

10.1 An Example Wind Turbine

We will analyze the behavior of a horizontal axis, three-bladed wind turbine example, as shown in Figure 10.1, with a rotor diameter of 100 m. The tower supporting the example turbine is shown in Figure 10.2. The tower is made of steel with 80 m height. Its diameter is 5 m at the bottom and 3.5 m at the top. The wall thickness is 40 mm at the bottom and 20 mm at the top (stepwise reduction is not accounted for). The foundation and the nacelle are assumed to be stiff.

A side view of the model of the complete structure is shown in Figure 10.3. This is an upwind turbine, as the rotor is located in front of the tower. The rotor is also tilted by 5 degrees in order to get a good clearance of the blade tip to the tower in case of wind gusts, a topic of Section 10.4.

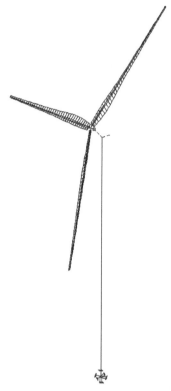

FIGURE 10.1
Wind turbine example with three blades.

10.2 Modeling and Analysis of Wind Turbine Blade

The velocity triangle for a wind turbine is shown in Figure 10.4. The difference between this and the triangle shown in Chapter 9, Figure 9.14 is the direction of power transfer with respect to the flow. The propeller blade is active in its acceleration of the airflow by putting energy into the system. The turbine blade is passive in the sense that it takes energy out of the system and decelerates the flow while doing so. These differences are manifested in the respective velocity triangles.

In the turbine case under consideration now, there is a tangential force (F) pointing forward on the profile that is driving the rotor. This force was pointing backward in the propeller case. The value of this force depends on the ratio of the tangential velocity at the tip with radius R and the wind velocity. This ratio is called the tip speed ratio and is defined as

$$\lambda = \frac{\Omega R}{w}.$$

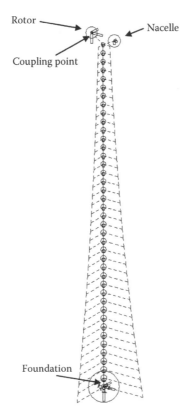

Rotor

Nacelle

Coupling point

Foundation

FIGURE 10.2
Tower model with stiffness, mass, and inertia.

This is the inverse of the advance ratio defined for the propeller, further demonstrating the distinction between the two cases.

When the wind is passing through the rotor, the effective wind velocity at the rotor plane is reduced by the so-called induced velocity. For wind turbines the induced velocity is represented by a factor applied to the free stream wind velocity w_0 as

$$w = w_0(1-a).$$

The induction factor can be found by considering the momentum of the rotor disk and the thrust found from the blade sections, as shown in Figure 10.4. Different methods are known for the calculation of the induction factor, for example, the method of Wilson and Lissamann [29].

The lift, drag, and moment on the blade sections are calculated in the same way as for the propeller in Chapter 9, Section 9.1. The aerodynamic coefficients for a typical wind turbine blade are shown in Figure 10.5 with

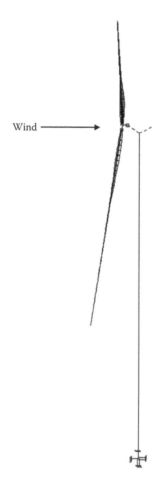

Wind ⟶

FIGURE 10.3
Complete finite element model of the example wind turbine.

linearized curves for the lift and the moment coefficients. The coefficients are given for the quarter-chord point. The resulting lift force is usually behind this point, which is accounted for by the moment coefficient.

Our example turbine is designed for a rated power of 3.7 MW at 12 m/s wind velocity and 20 RPM rotor speed. The power in watts is shown as a function of wind velocity in Figure 10.6. The power is a function of the cube of the wind velocity so the power increases nearly as a cubic parabola in the beginning. When the rated power is reached at the design wind speed of 12 m/s, the blades are pitched according to Figure 10.7 in order to limit the moment at the generator to the design value of 3.7 MW. Exceeding this design value can lead to failure of the generator.

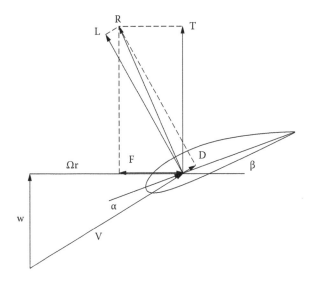

FIGURE 10.4
Velocity triangle for a wind turbine.

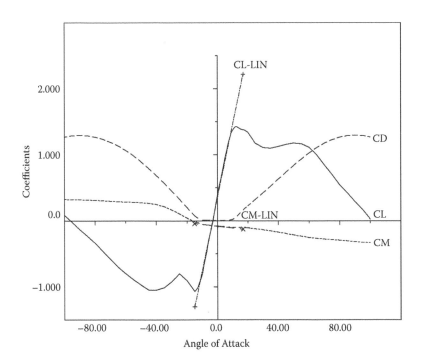

FIGURE 10.5
Aerodynamic coefficients for a typical wind turbine blade.

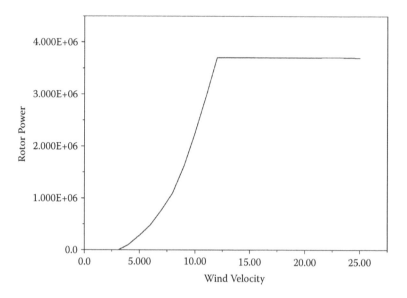

FIGURE 10.6
Rotor power as a function of wind velocity.

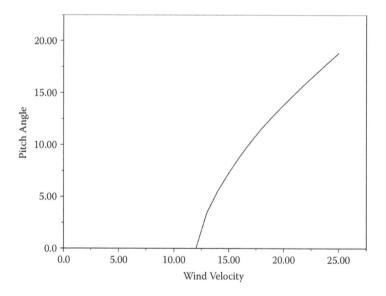

FIGURE 10.7
Blade pitch angle as a function of wind velocity.

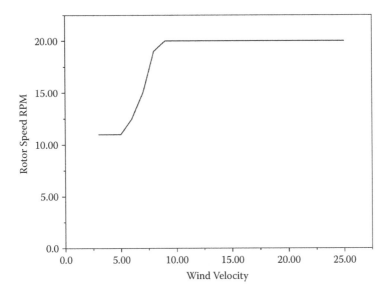

FIGURE 10.8
Rotor speed as a function of wind velocity.

At low wind speeds the rotor must also run slower in order to obtain a better aerodynamic efficiency. Once the wind velocity reaches the design point, the rotor speed is constant, as shown in Figure 10.8. The tip speed ratio as a function of the wind velocity is shown in Figure 10.9. After reaching the design velocity, it continues to decrease due to the constant rotor speed. The resulting

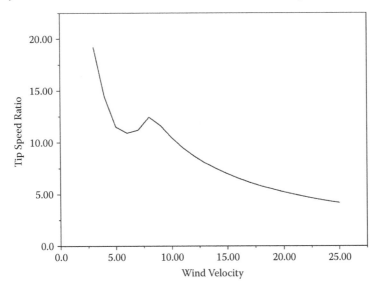

FIGURE 10.9
Tip speed ratio as a function of wind velocity.

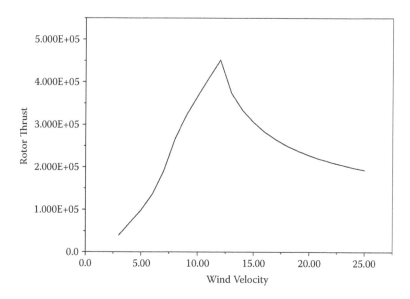

FIGURE 10.10
Rotor thrust as a function of wind velocity.

thrust of the rotor is shown in Figure 10.10. The maximum thrust of approximately 452,000 N occurs at a nominal operating condition at 12 m/s wind velocity. Before that, the rotor speed is lower, and above that, the blades are pitched and the thrust component in wind direction is reduced.

Let us now create an example blade for our turbine using fictive values of stiffness and mass as well as an assumed geometry. These values do not represent an existing turbine, but are realistic and adequate for the analysis demonstration. The stiffness distribution for the blade cross-sections as a function of the radius is shown in Figure 10.11.

The four curves indicate the stiffness with respect to out-of-plane (EI-flap) and in-plane (EI-lag) deformations and to the generic tension (EA) and torsion (GJ) stiffness values.

The mass and the radius of inertia about the local center of mass of the sections are shown in Figure 10.12. The chord length variation of the blade profile sections is shown in Figure 10.13.

As in the case of propellers, the wind turbine blades are also twisted in order to obtain an optimal angle of attack along the blade. The twist as a function of the radius is shown in Figure 10.14. A positive twist means that the blade is turned down as in Figure 10.4. The curve indicates that in the outermost region of the blade, between 40 and 50 m, the twist is negative.

To finalize the geometry of the blade, the offsets of the elastic axis (shear center), the center of mass, the centroid, and the quarter-chord lines are

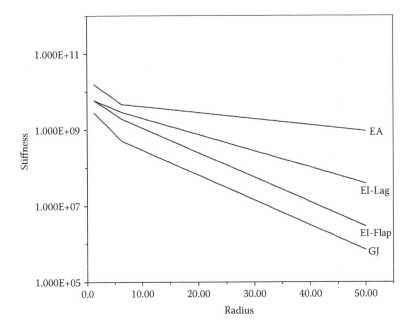

FIGURE 10.11
Stiffness values for the blade.

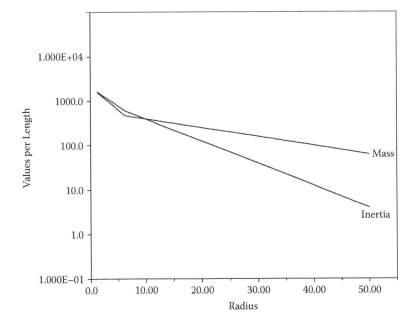

FIGURE 10.12
Mass and inertia per unit length.

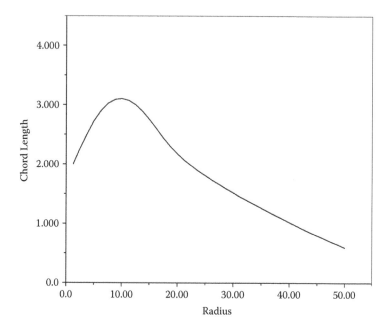

FIGURE 10.13
Chord length of the blade.

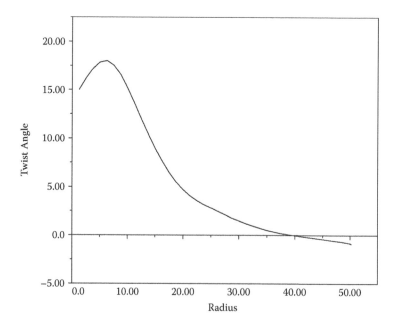

FIGURE 10.14
Blade twist angle.

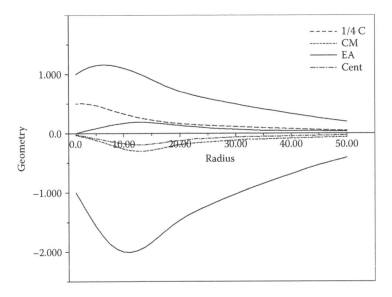

FIGURE 10.15
Offsets and blade geometry.

specified, as shown in Figure 10.15, together with the leading and trailing edge of the blade (in the figure, "CM" stands for center of mass and "Cent" is the centroid). All values are shown relative to the pitch axis. This is the reference axis, and for high wind speeds the blades are pitched about this axis.

The blade structure is discretized by beam finite elements. The beam elements are established between the nodes representing the elastic axis. The mass is distributed to two nodes with half the section's mass, and they are located at a distance equal to the radius of inertia, fore and aft of the center of mass. These nodes are connected to the shear center by means of stiff elements. In order to better see the blade, visualization elements (connections without stiffness) have been generated between the leading edge, the trailing edge, the quarter-chord points, and the centers of mass. The blade model is shown in Figure 10.16.

A detail of the blade model is shown in Figure 10.17. The aerodynamic elements are the sections between the chord-wise lines, and the aerodynamic forces are calculated for each of these. The nodes on the quarter-chord line and the leading edge are used to calculate the twist of the aerodynamic elements, and the aerodynamic forces and moments are applied to the quarter chord. The centers of the small circles are the masses.

A view along the blade axis, together with the wind direction and the direction of the air stream due to the tangential velocity, are shown in Figure 10.18. The figure shows the twisting of the blade.

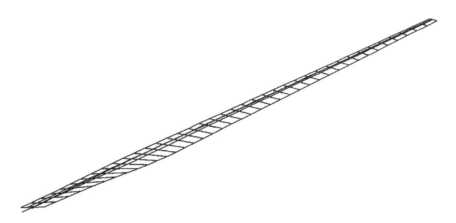

FIGURE 10.16
Finite element model of the blade.

Let us now analyze the blade alone. The blade is constrained at the hub, and first we compute the real eigenvalues and mode shapes. We recall from earlier discussions the four major mode shape types for blades. They are the flap modes that are blade bending perpendicular to the rotor plane, the lead-lag or (edgewise) modes that are bending mainly in the rotor plane, the torsion modes that are important for the stability of the blade, and the generalized force for the centrifugal force that is based on the extension

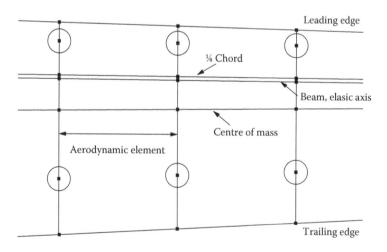

FIGURE 10.17
Detail of blade model.

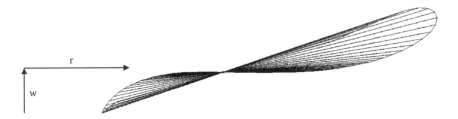

FIGURE 10.18
View along the blade axis.

(longitudinal) modes. These may be important in assessing the stresses occurring. Some of the important normal mode shapes of our example blade are listed in Table 10.1.

The first and second flap and lag modes, shown in Figures 10.19 through 10.22, are characteristic of regular cantilever beam mode shapes with some added effect caused by the twisted blade profile. The left-hand pictures show the view of the mode shapes in the plane of the rotation, and the right-hand pictures present a view perpendicular to that.

The first torsion mode (8) observed slightly above 8 Hz is shown in Figure 10.23.

TABLE 10.1

Modes of the Wind Turbine Blade

Mode	Eigenfrequency (Hz)	Generalized Mass (kgm²)	Shape
1	0.510	563.925	First flap
2	0.887	808.499	First lag
3	1.569	507.464	Second flap
4	2.978	670.329	Second lag
5	3.417	492.789	Third flap
6	6.031	483.861	Fourth flap
7	6.891	631.863	Third lag
8	8.074	100.742	First torsion
9	9.502	466.412	Fifth flap
10	12.502	649.663	Fourth lag
11	13.092	82.883	Second torsion
12	13.784	461.829	Sixth flap
18	26.369	2,475.948	First extension

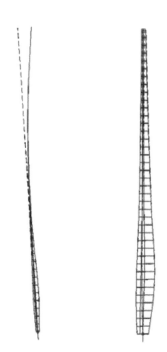

FIGURE 10.19
First flap mode.

Finally, the first extension mode (18), occurring at 26.4 Hz, is shown in Figure 10.24. Let us now turn to the rotor dynamic analysis of the example blade, including the quasi-steady aerodynamic forces. The eigenfrequencies are shown in Figure 10.25. The numbers correspond to the notations in Table 10.1 and the values on the right-hand side of the figure correspond to the values in the table.

The eigenfrequency of the blade torsion (mode 8) decreases due to the negative aerodynamic stiffness. The torsion mode will then couple with the closest bending mode, the lag mode (7) at around 27 RPM. Bifurcation occurs at 41 RPM where the eigenfrequency of the first flap mode goes to zero.

The real eigenvalues are shown in Figure 10.26. The real part of the solution after the bifurcation is slowly increasing, but it remains stable. In contrary, the real part of the propeller blade shown in Chapter 9, Figure 9.22 becomes unstable shortly after the bifurcation point. This is due to the fact that the damping of the flap modes of a wind turbine blade is much larger than that of the propeller. The damping factor is the ratio of the physical

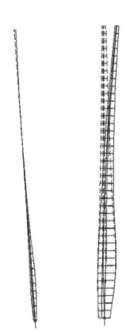

FIGURE 10.20
First lag mode.

damping due to the aerodynamic forces (*D*) and the product of the mass and the eigenfrequency as

$$\zeta = \frac{D}{2\omega M}.$$

The damping of the wind turbine blade, shown in Figure 10.27, is larger than that for the propeller blade, shown in Chapter 9, Figure 9.21. Note the difference in the scales between the two figures. For example, the damping for the first flap mode gets larger than one (100%) at 30 RPM and becomes super-critical. The damping of the torsion mode, however, is low.

The flutter instability starts around 30 RPM when mode 7 crosses the horizontal axis, which is well above the design (nominal) speed of 20 RPM for our turbine. However, during a gust, the rotor speed may increase significantly above the nominal speed. Therefore a certain safety margin must be considered in order to account for such uncertainties.

For an aircraft, this factor is related to the maximum diving speed, which is already higher than the maximum operational speed. For turbines, this factor is related to the maximum operation speed.

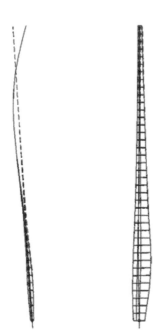

FIGURE 10.21
Second flap mode.

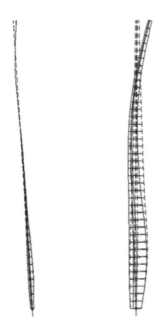

FIGURE 10.22
Second lag mode.

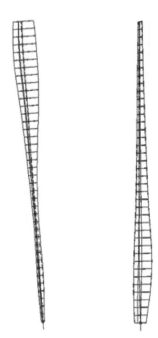

FIGURE 10.23
First torsion.

Other effects will also influence the flutter speed. When the rotor is running in icy conditions, the ice on the blade leads to a larger mass and moment of inertia. The eigenfrequencies will be decreased. Also the position of the center of mass may be changed and a lower critical speed can be expected.

In reality the rotor blades would start to pitch above 12 m/s wind velocity and the condition would be different. It is necessary to compare the critical rotor speed with the maximum rotor speed to evaluate the margin of safety.

This instability occurs due to the coupling between the torsion and the bending modes. The complex torsion mode at 40 RPM is shown in Figure 10.28. The flutter speed is influenced by the torsion eigenfrequency. Increasing the torsion stiffness leads to a higher critical speed.

The offset of the center of mass from the elastic axis causes a coupling between bending and torsion and has a strong influence on the stability behavior. The position of the center of mass as was shown in Figure 10.17 is far aft of the elastic axis.

The eigenfrequencies of an analysis without offset between the center of mass and the elastic axis is shown in Figure 10.29. Here the torsion mode is

FIGURE 10.24
First extension mode.

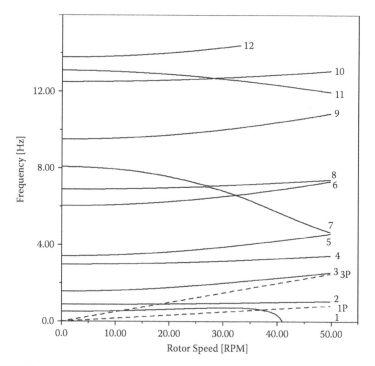

FIGURE 10.25
Eigenfrequencies for the blade.

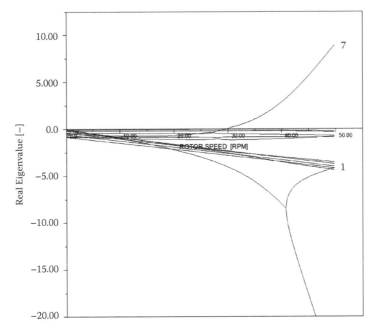

FIGURE 10.26
Real eigenvalues.

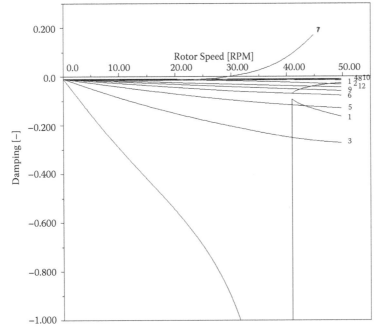

FIGURE 10.27
Damping of the blade.

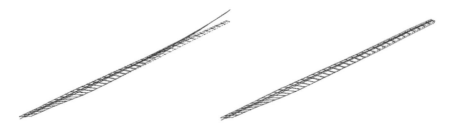

FIGURE 10.28
Real and imaginary part of the complex torsion mode at 40 RPM.

only slightly reduced and no bifurcation occurs in the calculated range of rotor speed.

The damping curves of the analysis without offset between the center of mass and the elastic axis are shown in Figure 10.30. It is clear that the blade is stable in the calculated speed range. In this case, extra calculation, above

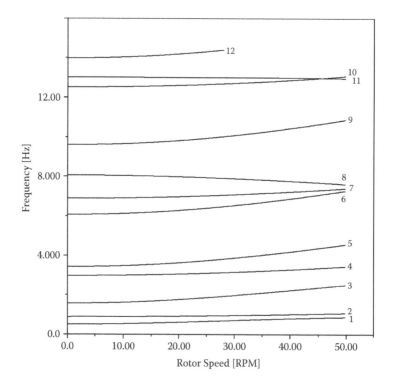

FIGURE 10.29
Eigenfrequencies for the blade without mass offset.

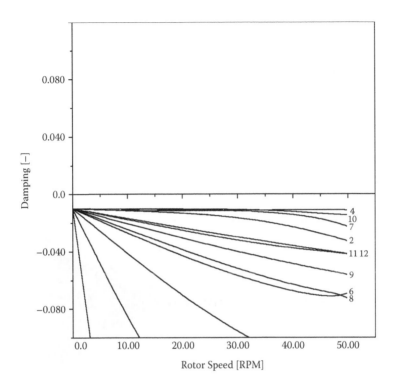

FIGURE 10.30
Damping curves for the blade without mass offset.

the computational rotor speed range of up to 50 RPM, showed the bifurcation to be at 80 RPM and the flutter instability to be at 100 RPM. These are not visible in the figure. Obviously, the mass offset has a great influence on the blade stability. This is also the case for the flutter behavior of aircraft. For both cases, a mass aft of the elastic axis has a detrimental effect on stability.

The instability can actually be verified by making a transient response analysis, the topic we discussed in generic terms in Chapter 7, Sections 7.3 and 7.4. We will use a starting point of 20 RPM and a constant wind velocity of 12 m/s and assume that the rotor speed and the wind velocity are linearly increased with time, as shown in the upper part of Figure 10.31. In order to trigger the instability, an impulse due to the tower influence is applied.

In the transient response analysis the aerodynamic forces are calculated directly at each time step using the instantaneous position, velocity, and acceleration of each aerodynamic element. Hence the aerodynamic mass, stiffness, and damping matrices are not applied on the left side of the equation. The result of this sweep analysis is shown in the lower part of Figure 10.31.

The torsional vibration of the tip gets unstable around 30 RPM, as shown by the projection line between the lower and upper parts of the figure. This

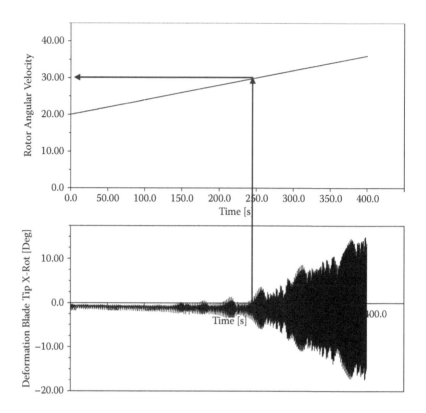

FIGURE 10.31
Torsion of the blade tip resulting from a sweep analysis.

is in good agreement with the results found in the eigenvalue analysis shown in Figures 10.26 and 10.27, where the torsion instability also occurred at 30 RPM. In the transient analysis the variable aerodynamic coefficients according to Figure 10.5 have been used, not the linearized curves, and the induction has also been accounted for.

Finally, we analyze our turbine blade with the inclusion of the unsteady aerodynamic forces and the solution is obtained by the PK iteration method described in Chapter 9, Section 9.3. The eigenfrequencies are shown in Figure 10.31. The results are similar to the analysis with quasi-steady aerodynamic forces shown in Figure 10.25. In the quasi-steady case the solution line numbers 7 and 8 were not crossing, albeit they coupled, and in Figure 10.30 they are crossing.

The damping curves are shown in Figure 10.33. Now the flutter instability speed has increased to 44 RPM, as shown by the crossing of mode 8. The reason for this is the reduction of the aerodynamic stiffness due to the real part of

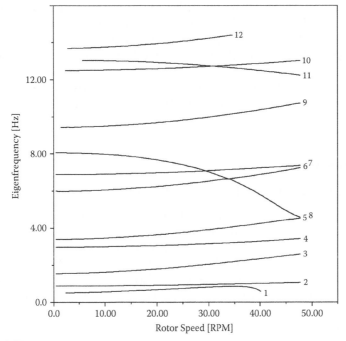

FIGURE 10.32
Blade eigenfrequencies with unsteady aerodynamic forces.

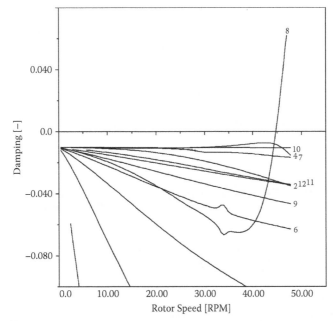

FIGURE 10.33
Damping curves for blade with unsteady aerodynamic forces.

the Theodorsen function and the different phase due to the imaginary part. This has also been found by Lobitz [30] comparing the quasi-steady and unsteady aerodynamic forces for large wind turbine blades. It was also found that for large blades the flutter instability gets closer to the operating speed.

We have exhausted the practical analysis scenarios of the stand-alone blade. We will now turn our attention to the complete turbine.

10.3 Wind Turbine with Three Blades

The example turbine with three blades was shown in Section 10.1. We assumed that the hub and the main shaft were stiff with a high Young's modules and the drive train with gear box, couplings, and generator were not modeled. Because of the stiff hub, shaft, and drive train the eigenfrequencies are practically equal to those of the blade, but now there are three distinct mode types per blade mode. There is a symmetric mode shape where all blades are moving in the same direction, and two antisymmetric modes where the blades are moving in opposite direction. This behavior was also encountered in the example of four arms in Chapter 6, Section 6.8.

Thus far we omitted the elastic deformation of the blades due to the steady load. As we saw in Figure 10.10 the rotor thrust is around 450,000 N at the design point of 12-m/s wind velocity and 20-RPM rotor speed. Therefore at this condition the bending of the blade will be significant. The forces on the blade are shown in Figure 10.34. The upper picture shows the aerodynamic forces and the lower picture shows the centrifugal forces.

Applying these forces to the model and executing a static nonlinear analysis, the static displacement of Figure 10.35 is obtained. The deformation at the tip is 5.48 m in the wind direction and against the coning angle. This means that the blade will be bent toward the tower. The tip position due to the coning angle of

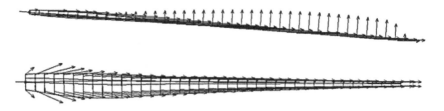

FIGURE 10.34
Aerodynamic and centrifugal forces on the blade.

FIGURE 10.35
Static deformation of the blade.

4 degrees is around 3.5 m. Hence the elastic deformation is almost 2 m larger due to the coning angle.

The deformation found from this nonlinear analysis with 10 load increments is now compared to the deformation of the blade tip found in a transient response analysis, using the geometric stiffness matrix calculated for unit rotor speed and without updating the geometric stiffness at every step. The displacement of the blade tip is shown in Figure 10.36. The oscillations are due to the gravity excitation of the rotor. The mean value is 5.22 m, which is only 4.7% lower than the value found by nonlinear static analysis.

This elastic deformation will lead to a stronger Coriolis coupling between flap and lag motions. In addition to this, there will be a kinematic coupling between lag and torsion.

We now execute a rotor dynamic analysis by including this blade deformation and the geometric stiffness. For example, the second lag mode shape of this analysis is shown in Figure 10.37, demonstrating the torsion of the blade tip.

The eigenfrequency results of this analysis are shown in Figure 10.38, showing only the symmetric modes. The flap modes are almost unchanged compared to those of Figure 10.25, but the lag and torsion modes are different.

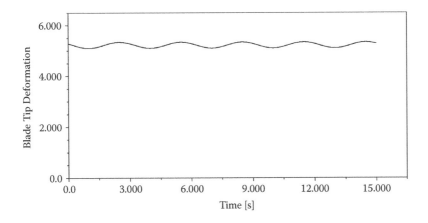

FIGURE 10.36
Blade tip deformation calculated in a transient analysis.

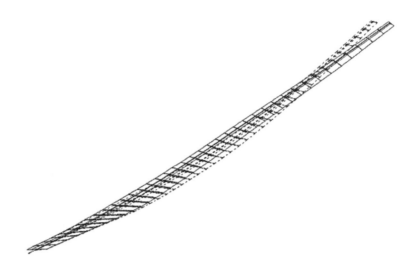

FIGURE 10.37
Second lag mode with torsion.

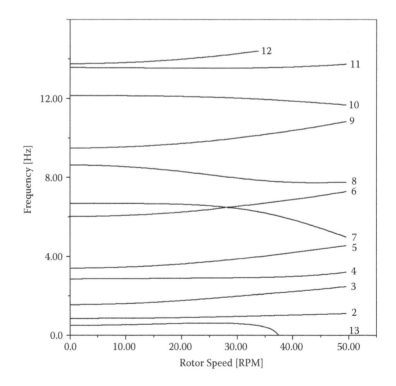

FIGURE 10.38
Eigenfrequencies with static deformation.

TABLE 10.2

Mode Shapes of the Wind Turbine Tower

Mode	Eigenfrequency (Hz)	Generalized Mass (kgm²)	Shape
1	0.434	180,069.6	Lateral bending
2	0.438	176,664.1	Longitudinal bending
3	1.762	1,012,810.2	Torsion
4	2.076	302,572.8	Nacelle roll
5	2.566	161,315.3	Nacelle pitch
6	3.217	7,063,448.0	Torsion
7	4.236	148,584.7	Lateral bending
8	5.751	181,668.6	Longitudinal bending
9	9.791	99,812.7	Lateral bending
10	10.833	214,301.9	Vertical extension
11	11.447	119,026.4	Longitudinal bending

It is noticeable that the static (0 RPM) eigenfrequencies are higher than those in Table 10.2. There is again a coupling between modes 8 and 7, noticeable by their respective inflections (around the point where 6 and 7 cross), but not in such close proximity as earlier.

The damping curves are shown in Figure 10.39. The instability mode (7) remains around 32 RPM, but now there is a strong instability for the second lag mode (4) around 25 RPM and a weaker instability of the first lag mode (2) at 17 RPM. These new instabilities are mainly due to the coupling of lag and torsion of the blade's bending under the static forces.

We computed the geometric stiffness matrix only due to the centrifugal force. Equation (10.1) demonstrates this for a beam element, where T is the tension force arising from the centrifugal force and L is the element length.

$$K_{G1} = \frac{T}{L} \begin{bmatrix} 0 & 0 & 0 & 0 & 0 & 0 \\ 0 & \dfrac{6}{5} & 0 & 0 & 0 & -\dfrac{L}{10} \\ 0 & 0 & \dfrac{6}{5} & 0 & \dfrac{L}{10} & 0 \\ 0 & 0 & 0 & 0 & 0 & 0 \\ 0 & 0 & \dfrac{L}{10} & 0 & \dfrac{2L^2}{15} & 0 \\ 0 & -\dfrac{L}{10} & 0 & 0 & 0 & \dfrac{2L^2}{15} \end{bmatrix} \tag{10.1}$$

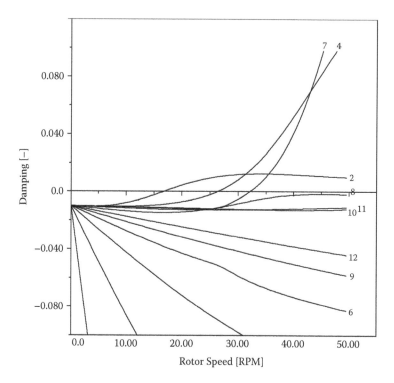

FIGURE 10.39
Damping curves with static deformation.

Furthermore, there are also terms due to element bending moments (M_y) and M_z) and shear forces (F_y and F_z), as shown in Equation (10.2). For a heavily loaded wind turbine blade the latter terms can also be very important

$$
K_{G2} = \begin{bmatrix}
0 & 0 & 0 & 0 & 0 & 0 \\
0 & 0 & 0 & \dfrac{M_y}{L} & 0 & 0 \\
0 & 0 & 0 & \dfrac{M_z}{L} & 0 & 0 \\
0 & \dfrac{M_y}{L} & \dfrac{M_z}{L} & 0 & \dfrac{F_y}{6} & \dfrac{F_z}{6} \\
0 & 0 & 0 & \dfrac{F_y}{6} & 0 & 0 \\
0 & 0 & 0 & \dfrac{F_z}{6} & 0 & 0
\end{bmatrix}
\tag{10.2}
$$

The centrifugal forces were multiplied by Ω^2 in the rotor dynamics equilibrium equations. Let us now consider the steady aerodynamic forces. The lift on a blade segment is

$$L = \frac{\rho}{2} V^2 c s C_L. \tag{10.3}$$

The square of the total velocity is found from the velocity triangle in Figure 10.4 as

$$V^2 = \Omega^2 r^2 + w^2. \tag{10.4}$$

With the dimensionless radius

$$\eta = \frac{r}{R}$$

and the tip speed ratio

$$\lambda = \frac{\Omega R}{w},$$

the velocity may be expressed as a function of rotor speed and geometry:

$$V^2 = \Omega^2 R^2 (\eta^2 + \lambda^2). \tag{10.5}$$

Hence the loads due to the aerodynamic forces are also a function of the square of the rotor speed when the tip speed ratio is kept constant. Therefore the geometric stiffness matrix can be calculated for the elements due to both centrifugal forces and aerodynamic forces calculated for a unit rotor speed.

Applying the full geometric stiffness without the static deformation for our example rotor blade, we obtain the frequency results of Figure 10.40. There is again the coupling relationship between modes 7 and 8.

The damping curves in Figure 10.41 clarify the scenario. Mode 7 again became unstable; however, there is an additional instability of the second lag mode (4) coupled with torsion. Because the eigenfrequency of the lag mode is lower than that of the torsion, the reduced frequency is lower. Hence the influence of the Theodorsen function is smaller, and the lag instability is less influenced by the unsteady aerodynamics as the torsion instability.

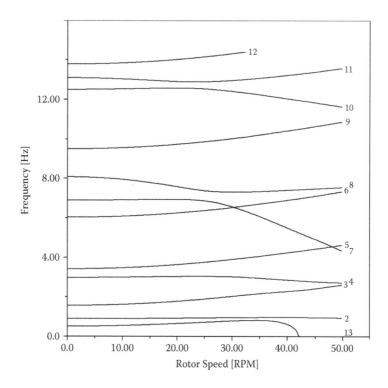

FIGURE 10.40
Eigenfrequencies for the blade with the full geometric stiffness matrix.

The differences between the geometric stiffness based only on the centrifugal force as opposed to being based on all the forces are demonstrated in Figure 10.42. The dashed lines are the values with only centrifugal force and the solid lines represent the geometric stiffness matrix with all forces. The diagonal terms of the generalized geometric stiffness matrix are almost equal except for the second torsion and the fourth lag modes, as shown in Figure 10.42.

The reason for the different behavior is the addition of the shear forces and bending moments of the blade element and the coupling with the blade torsion in the geometric stiffness matrix. These terms occur only when the blade is loaded by the aerodynamic forces. The pattern of the geometric stiffness matrix with only the centrifugal forces is shown in Figure 10.43 and the case of all forces in Figure 10.44. In the first case the dominant terms are on the diagonal, and in the latter case there are large off-diagonal coupling terms as well.

Finally, let us include both the full geometric stiffness matrix and the static deformation into the analysis of the blade. The eigenfrequency results are

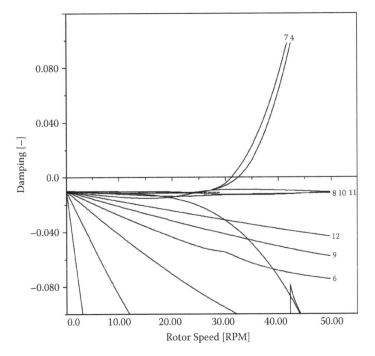

FIGURE 10.41
Damping curves for the blade with full geometric stiffness matrix.

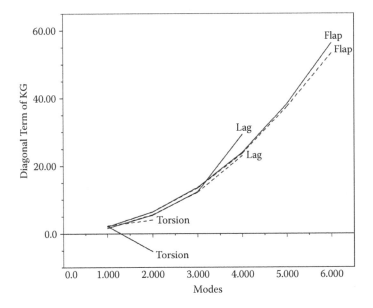

FIGURE 10.42
Diagonal terms of the geometric stiffness matrix.

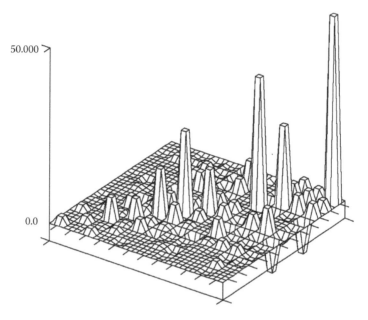

FIGURE 10.43
Generalized geometric stiffness matrix with only centrifugal forces.

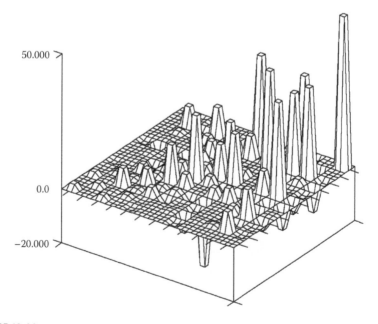

FIGURE 10.44
Generalized geometric stiffness matrix with all forces.

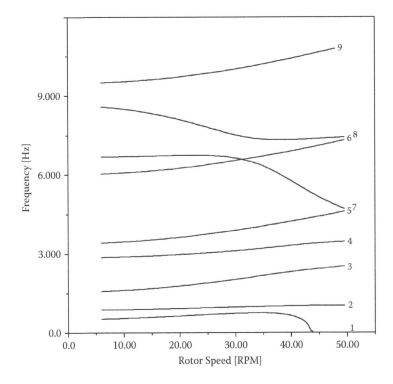

FIGURE 10.45
Eigenfrequencies with static deformation and full geometric stiffness.

shown in Figure 10.45. The earlier tendencies are preserved. Figure 10.46 presents the damping curves. Now the two lag modes that became unstable earlier are damped. The damping of the first lag mode (2) is reduced around 20 RPM and the damping of the second lag mode (4) is close to zero around 38 RPM. The torsion instability remained, now at slightly above 31 RPM.

In summary, in the case of the undeformed blade the full geometric stiffness matrix caused a lag instability. The statically deformed blade, however, became more stable when the full geometric stiffness matrix was used.

In practice, wind turbines must be analyzed for many conditions with different wind velocities and pitch angles in order to get an overview of the whole operating range. For certain wind velocities the flap and lag modes may be turned—a phenomenon that influences the stability. Furthermore, different steady force distributions and static deformations may lead to lower critical rotor speeds. These topics are not explored here further.

We have not yet considered the tower's influence in the behavior of the blades of our example turbine. First, we compute the eigenfrequencies and

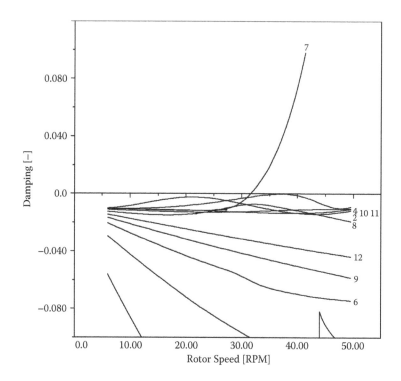

FIGURE 10.46
Damping with static deformation and full geometric stiffness.

mode shapes for the tower, including the rotor mass and inertia. These mass and inertia values are then subtracted before the mass of each point of the rotor structure is added. The eigenvalues and modes are listed in Table 10.2. All the listed mode shapes are shown in Figures 10.47 and 10.48.

When coupling with the rotor structure, modes 1, 2, and 10 are mainly acting as translations of the rotor as described in Chapter 4, Section 4.4. Modes 3, 4, 5, and 6 will rotate, the rotor, and therefore the fully coupled equations of Chapter 2, Section 2.4 will apply. The time-dependent terms are implemented in the form of Chapter 2, Equations (2.34) through (2.37). The aerodynamic matrices are defined in a similar fashion.

In the coupled analysis of our example turbine, we used the modal solution method described in Chapter 4, Section 4.4, and the modal displacements and rotation of the coupling point with the tower were used in the equations. In order to check this modal coupling, we compared the results to those found in the direct rotor dynamic analysis (Chapter 4, Section 4.3) at zero rotor speed. The eigenfrequencies for the first 48 modes are shown in Figure 10.49. In the modal coupling analysis 36 rotor and 14 tower modes were accounted for. There is a visibly very good agreement between the direct and modal methods in the modes that are shown.

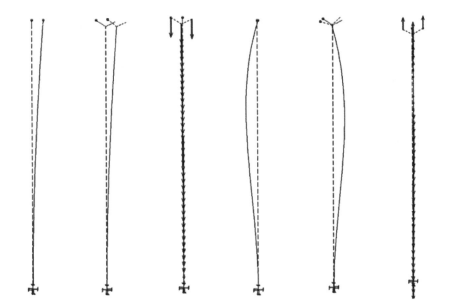

FIGURE 10.47
Tower modes 1 through 6.

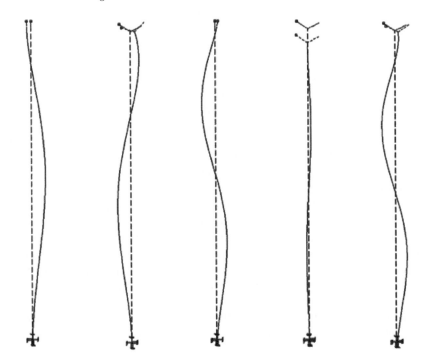

FIGURE 10.48
Tower modes 7 through 11.

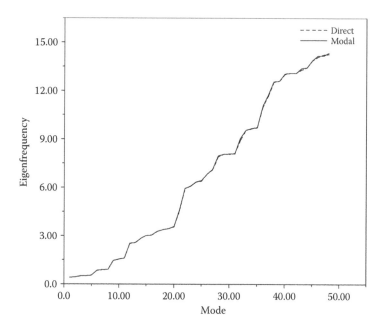

FIGURE 10.49
Comparison of eigenfrequencies with the modal and direct methods.

We also executed a stability analysis for the coupled structure. The eigenfrequencies are shown in Figure 10.50. The mode numbers are shown for clarity only for the solutions that become unstable. There is a noticeable crossing of the 3P line with the lag modes around 20 RPM that can lead to very undesirable resonance phenomena between the tower and the rotor.

The damping curves are shown in Figure 10.51. The three strong instabilities are due to blade torsion and are similar to the instability of the blade shown in Figure 10.27. The complex mode of the symmetric blade torsion is shown in Figure 10.52. This is the mode number 29 in Figure 10.51.

In addition to the three torsion modes (22, 26, and 29) there is a new instability (30) starting already at 20 RPM. This is due to a coupling between a tower mode and blade torsion and lag. The complex mode is shown in Figure 10.53. If this were to occur in practice, we would have to make modifications to the blade in order to avoid this instability. The mode would need further analysis, including the aerodynamic forces in the geometric stiffness matrix and also the static deformation.

Finally, we must consider the fact that the blades are rotating and the coupling relationship changes as a function of the azimuth angle of the blade.

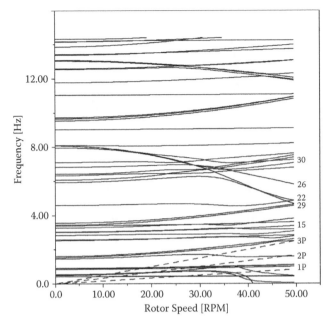

FIGURE 10.50
Eigenfrequencies of the coupled structure.

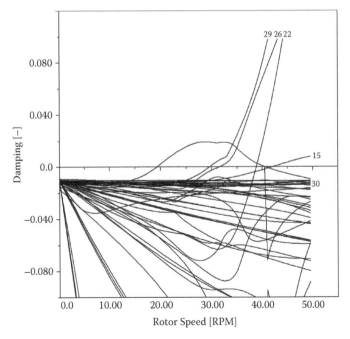

FIGURE 10.51
Damping curves for the coupled structure.

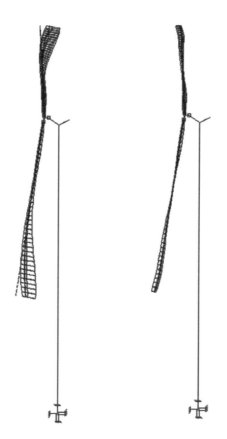

FIGURE 10.52
Real (left) and imaginary parts (right) of symmetric blade torsion, curve 29.

Even though the equation system describing this has periodic terms, the solution is not periodic. Instead of calculating the rotor with a fixed azimuth angle and with variable rotor speed, a "snap-shot" analysis can be done with a fixed rotor speed and with a variable azimuth angle. The eigenfrequencies for such an analysis of our example turbine (including the aerodynamic matrices) are shown in Figure 10.54. The frequencies are practically constant over the 360-degree rotor angle.

The damping curves are shown in Figure 10.55. These curves are also nearly constant over the rotor angle.

The slight variations for some modes may be due to the aerodynamic forces influencing the numerical accuracy of the computations. However, the welcome conclusion is that the coupled eigenvalue equations can be solved with the classical (nonperiodic) methods.

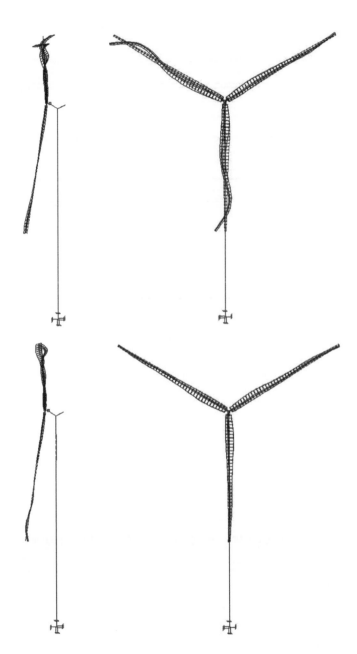

FIGURE 10.53
Real (upper) and imaginary parts (lower) of unstable solution 30.

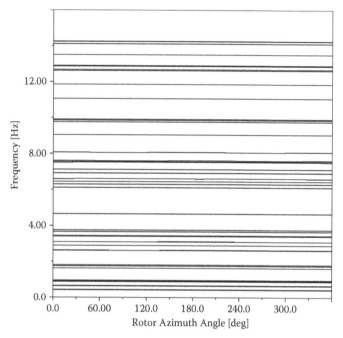

FIGURE 10.54
Eigenfrequencies as a function of azimuth angle for 20 RPM rotor speed.

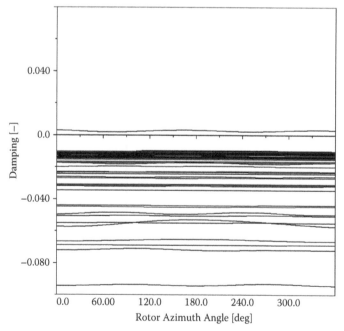

FIGURE 10.55
Damping as a function of azimuth angle for 20 RPM rotor speed.

10.4 Response Analysis of Wind Turbines

In this section we consider the response analysis of our example wind turbine. These analyses are mostly done in the time domain, but in some cases Fourier transformation is employed to review results in the frequency domain where it is easier to detect resonances. Response analysis of wind turbines considers several source of excitation forces:

1. Wind gradient. The wind velocity is considered to be zero at the ground and increases with height using an exponential relation.

2. Tower influence. If the rotor is located behind the tower (downwind), there is a strong wake induced by the tower and the wind velocity is decreased. For upwind rotors, the flow velocity is reduced due to the stagnation point of the tower.

3. Coning and tilt of the rotor. The wind component perpendicular to the blade is different at the various locations of the blade.

4. Gravity force. It gives rise to a strong bending moment variation in the lag direction of the blades.

5. Wind gusts. They result in large flap deformations that may require increasing the tower clearance.

6. Mass unbalance. This could be due to loss of ice on blades and produces strong torsional deformations.

The wind gradient is given by the formula

$$w_0 = w_H \left(\frac{h}{H} \right)^\alpha .$$

Here w_H is the reference wind velocity at the hub, h is the actual height, and H is the height of the hub. The exponent of the wind gradient is α.[*] A typical value for a strong wind gradient is 0.2. The resulting wind velocity profile is shown in Figure 10.56.

The tower influence may be approximated by

$$w = w_0 \left(1 - f \cos^2 \pi \frac{y}{b} \right).$$

Here y is the lateral distance from the center of the tower, b is the width of the region, and f is the reduction factor. For an upwind rotor (such as our example turbine) the speed reduction is smaller. In both upwind and downwind cases the tower influence is limited to a small region around the tower.

---

[*] This is not the angle of attack, but we use this in adherence to industry standards.

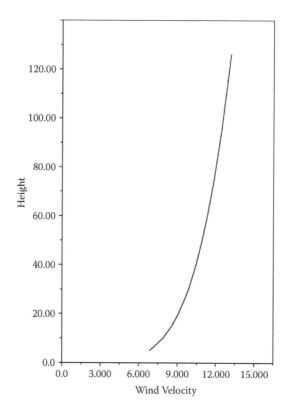

FIGURE 10.56
Wind gradient.

An example of using a reduction factor of 0.2 is shown in Figure 10.57. This leads to a tower influence factor of 0.8. We executed a steady state simulation of the example turbine with a tower influence factor of 0.9 (reduction of 0.1) and an exponent of wind gradient $\alpha = 0.2$. The initial conditions were zero, as in the simulations in Chapter 7, Section 7.4. Before the actual steady-state solution is attained, stabilizing runs had to be performed until the unsteady parts of the solution vanished.

The results of this simulation are shown in the following. For this type of analysis it is very useful to plot the results over the rotor azimuth angle. Figure 10.58 shows the inflow velocity for an element on a particular blade: blade 1. The azimuth angle is zero when blade 1 is pointing down. The lag deformation, shown in Figure 10.59, is mainly due to the gravity force, which acts as a once-per-revolution force. In addition there is a static component due to the tangential force that drives the rotor. The maximum lag deformation value occurs at 270 degrees, and the minimum value occurs at 90 degrees rotor angle.

The flap deformation is shown in Figure 10.60. The largest deformation occurs around 45 degrees after the top position, or around 225 degrees, where the wind velocity is highest. This is a 1P excitation force with a large

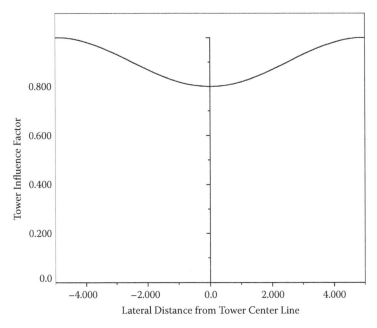

FIGURE 10.57
Tower influence factor.

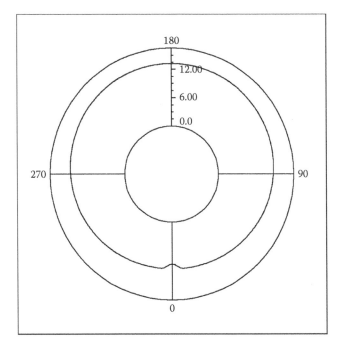

FIGURE 10.58
Inflow velocity for blade 1.

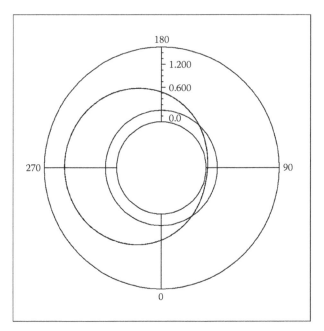

FIGURE 10.59
Lag deformation at the tip of blade 1.

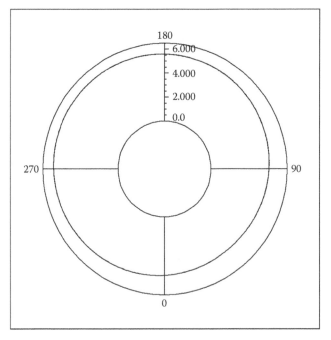

FIGURE 10.60
Flap deformation at the tip of blade 1.

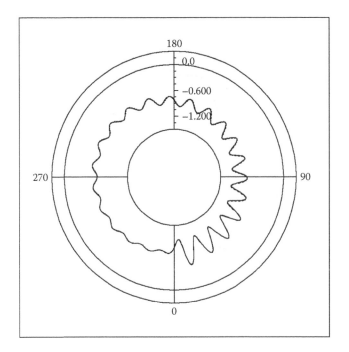

FIGURE 10.61
Torsion deformation at the tip of blade 1.

static part due to the thrust. The aerodynamic damping of the blade flap bending is large.

The blade tip torsion shown in Figure 10.61 is excited by the tower disturbance. The tip speed of the blade is around 100 m/s and the width of the tower disturbance is 10 m; the impulse is 0.1 s. The torsion eigenfrequency is 8.1 Hz and the aerodynamic damping is low. Therefore we see a large amplitude when the blade passes the tower disturbance (around 0 degrees) and then the amplitudes slowly decrease until the next impulse occurs.

The torsion moment at the bottom of the tower is shown in Figure 10.62. Because the rotor has three blades we see a pronounced 3P amplitude pattern, but there are also higher harmonics.

The aerodynamic power is found directly from the aerodynamic forces. The function is shown in Figure 10.63 with a noticeable disturbance for each blade. The mean value is lower than the design value of 3.7 MW because we have considered a strong wind gradient in the analysis.

There is also a relatively strong blade torsion, as shown in Figure 10.64. This has the effect of decreasing the pitching angle and the power is further reduced. The elastic torsion of the blade can be important and should be accounted for in the active control system of wind turbines.

The reason for the static torsion is the shape of the elastic axis shown in Figure 10.15.

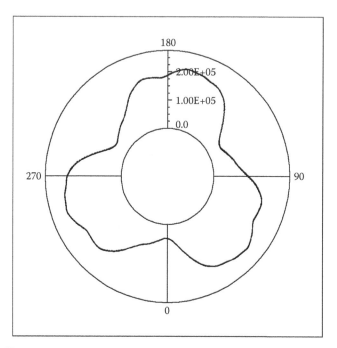

FIGURE 10.62
Torsion moment at the bottom of the tower.

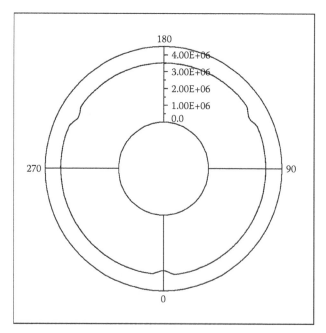

FIGURE 10.63
Rotor aerodynamic power.

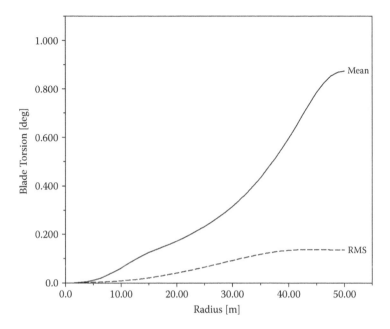

FIGURE 10.64
Mean value and RMS value of the torsion along the blade.

In another simulation scenario the rotor shaft is left free and a generator moment is applied to act against the aerodynamic torque from the rotor. Here a simple linear relationship between generator moment and rotor speed is used. The result is an acceleration matrix in the rotating part that represents the acceleration moment about the rotor axis.

When we consider a gust of 12 m/s added to the steady wind velocity, the total velocity curve of Figure 10.65 is obtained. The assumed gust starts at 2 s after start of the simulation. The duration of the gust is 5 s. The gust shape is a simple cosine function with a period of T_G.

$$w_G = \frac{w}{2}\left(1 - \cos 2\pi \frac{t}{T_G}\right).$$

Let our rotor be accelerated by the above gust. The maximum speed of 24.4 RPM is reached at the end of the gust, as shown in Figure 10.66. The aerodynamic power also strongly increases, as shown in Figure 10.67, but the power on the generator, as shown in Figure 10.68, increases to a lesser extent because the large inertia of the rotor is accelerated. After the gust has passed, the kinetic energy of the rotor is large and the rotor speed decreases slowly.

The forces on the blades due to the gust are large and the blade tip of our example turbine is bending 7.9 m toward the tower, as shown in Figure 10.69.

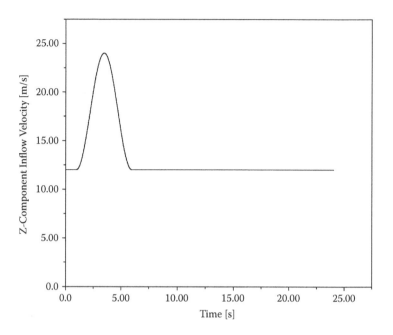

FIGURE 10.65
Total wind velocity at the hub due to gust.

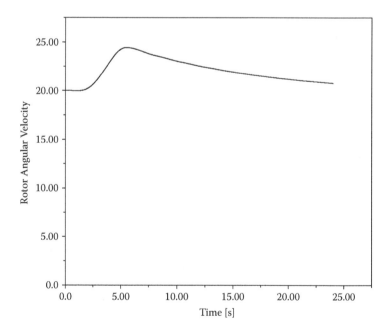

FIGURE 10.66
Rotor speed due to gust.

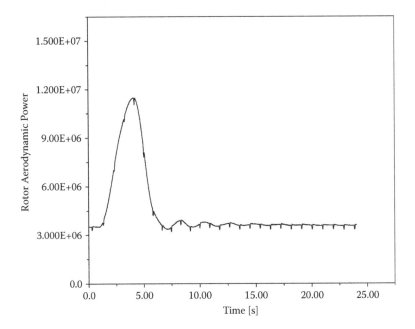

FIGURE 10.67
Aerodynamic power due to gust.

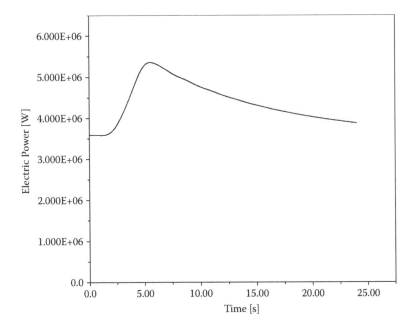

FIGURE 10.68
Generator power due to gust.

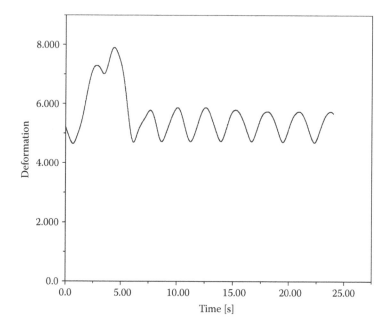

FIGURE 10.69
Blade tip flap deformation due to gust.

With a coning angle of 4 degrees, a tilt angle of 5 degrees, and a 4.25-m offset of the hub, the distance to the tower center line is around 11.75 m. The tower radius is around 2.2 m at the position of the blade tip. This results in a clearance of only 1.6 m at the blade tip. Based on this consideration our example turbine design should be modified.

A mass unbalance scenario occurs in wind turbines, for example, when ice builds up on the blades resulting in an additional mass accumulated on each blade. The total mass of a blade in our example model is around 13.4 tons. The area of the blade is around 90 m^2. Considering an ice thickness of 1 cm on both sides and a density of ice around 800 kg/m^3, we would end up with 1.4 tons on each blade.

We executed a simulation by distributing 1000 kg of extra mass on each blade. A transient response was done in order to obtain a steady solution. Then the ice on one blade was removed and the equation system was built up with the unsymmetric model. A new analysis was then executed with the results of the first simulation as initial conditions.

Due to the different centrifugal forces there is now a strong unbalance. When the ice was suddenly lost there was a large blade tip flap deformation, as shown in Figure 10.70. First the blade bent 12 m away from the tower and then 9 m toward the tower.

The blade torsion was also strongly excited because it was assumed that the ice was suddenly lost. Figure 10.71 shows the blade tip torsional deformation.

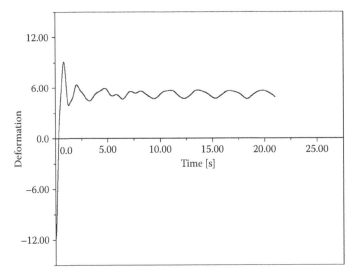

FIGURE 10.70
Blade tip flap deformation due to mass unbalance.

The tower sideways deformation is shown in Figure 10.72. The amplitudes are almost 0.4 m. The bending moments in wind and perpendicular to the wind direction are shown in Figures 10.73 and 10.74, respectively. Finally, the tower torsion moment is shown in Figure 10.75. All of these deformations present high amplitudes to be concerned about.

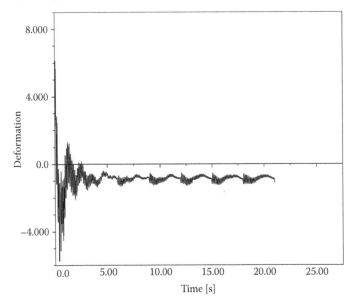

FIGURE 10.71
Blade tip torsional deformation.

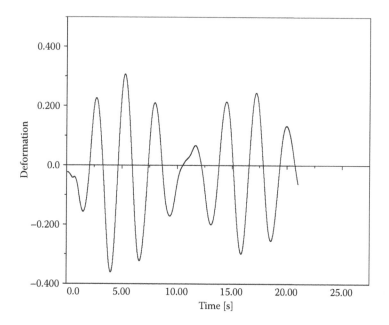

FIGURE 10.72
Tower sidewise displacement.

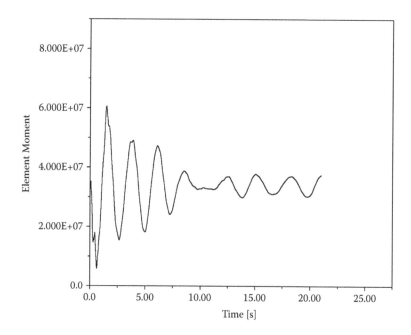

FIGURE 10.73
Tower bending moment in the wind direction.

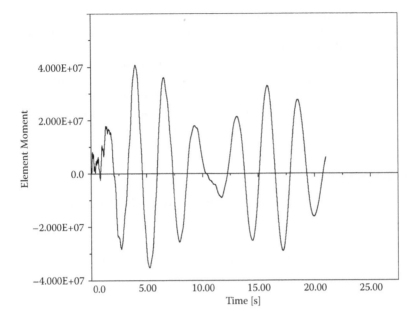

FIGURE 10.74
Tower sidewise bending moment perpendicular to the wind.

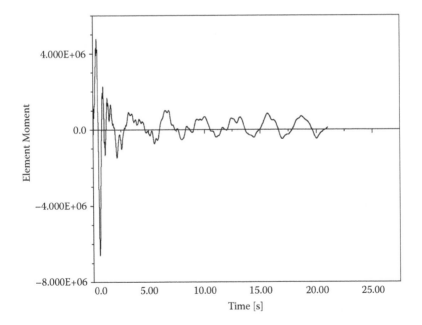

FIGURE 10.75
Tower torsion.

The mass unbalance due to sudden loss of ice cover is an extremely harsh operational scenario and must be carefully considered in wind turbines installed in colder climates.

For a wind turbine with three blades, the sine and cosine terms of the periodic matrices of the coupled formulation developed in Chapter 2, Section 2.5, cancel out, and we use the regular complex eigenvalue analysis for the solution with the rotor in any position. This is not the case with two-bladed wind turbines, the topic of our concluding section.

10.5 Horizontal Axis Wind Turbines with Two Blades

Two-bladed wind turbines are also built, albeit less frequently. Such structures are easier to assemble and two blades are certainly cheaper than three. However, their rotor dynamic analysis is much more complex. A wind turbine with two blades is shown in Figure 10.76.

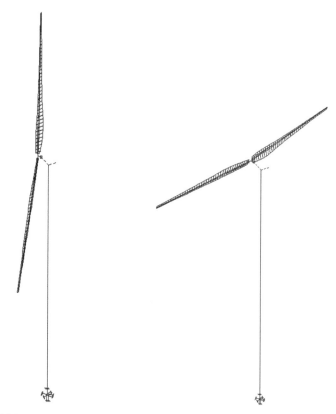

FIGURE 10.76
Wind turbine with two blades in vertical and horizontal positions.

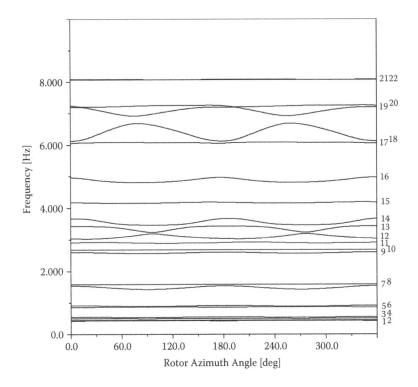

FIGURE 10.77
Eigenfrequencies of the coupled structure without rotation.

A two-bladed rotor would have blades with larger chord and thickness due to both aerodynamic efficiency and structural load considerations, because the two blades must take the whole thrust force.

In this section we will use our example turbine, with the same tower structure as for the three-bladed case, but now with only two of the same blades.

The coupling with the tower is different when the blades are in horizontal and in vertical positions. Therefore the eigenfrequencies (without considering rotation) are dependent on the rotor position, as shown in Figure 10.77. There are certain modes (for example, 18, 19) that express a pronounced periodic change as a function of the azimuth angle.

Let us further investigate the above scenario and calculate the eigenvalues for different rotor positions but at a certain, constant rotor speed. The damping curves for 18 RPM rotor speed are shown in Figure 10.78. Here the mode number 22 demonstrates a periodic behavior that is partly stable (damping is negative) and partly unstable (damping is positive).

The overall stability of that mode may be evaluated by integrating the curve over the 360-degree rotor angle at a certain rotational speed and observing the last value. The integral for 18 RPM is shown in Figure 10.79. Here the last value is still negative and the mode is stable at that speed.

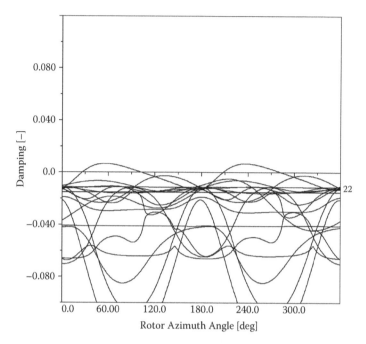

FIGURE 10.78
Damping as a function of the rotor angle.

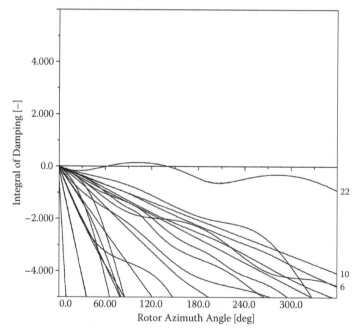

FIGURE 10.79
Integral of the damping at 18 RPM.

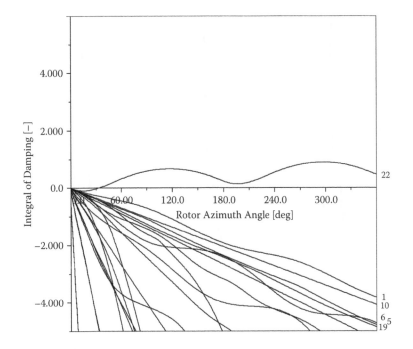

FIGURE 10.80
Integral of the damping at 20 RPM.

A similar plot is shown for 20 RPM in Figure 10.80. Here the integral value is positive and the mode is unstable at that speed. We would therefore expect the instability to arise between them.

The unstable mode 22 is the antisymmetric blade torsion, as shown in Figure 10.81, for the rotor in the vertical position. The real part is blade torsion, demonstrating the antisymmetric deformation of the blades, and the imaginary part is mainly blade bending. The kinetic energy of the mode is shown in the right picture and demonstrates the significant torsion of the tower. The same mode with the rotor in the horizontal position is shown in Figure 10.82.

The eigenfrequencies for 30 RPM are shown in Figure 10.83. Now the periodicity is even stronger than the case without rotation shown in Figure 10.77. There are now two unstable modes, as shown in Figure 10.84. The newly emerged instability is the symmetric torsion shown in Figure 10.85.

The symmetric blade torsion is already unstable at 30 RPM, according to this simple snap-shot method. The instability speed of 30 RPM was found for the blade alone and was already shown in Figure 10.26. The symmetric blade torsion for the three-bladed rotor also becomes unstable at 30 RPM, as was shown in Figure 10.51. This engineering snap-shot method is not based on theory, but shows clearly the periodicity of the system. The stability border, however, cannot be calculated precisely.

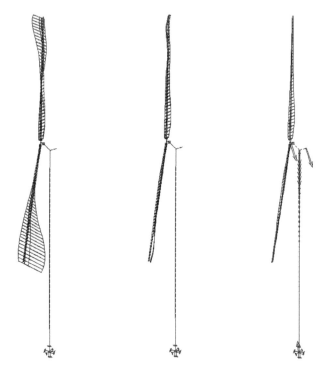

FIGURE 10.81
Real and imaginary parts, and kinetic energy of the unstable mode at 20 RPM—vertical.

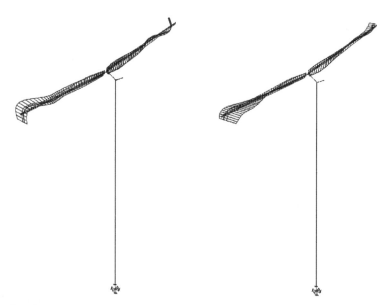

FIGURE 10.82
Real and imaginary parts of the antisymmetric blade torsion at 20 RPM—horizontal.

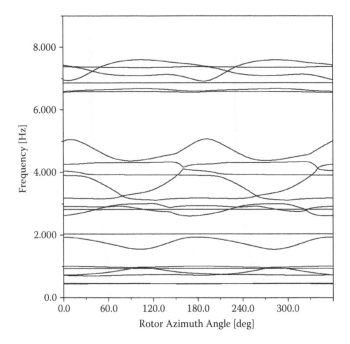

FIGURE 10.83
Eigenfrequencies as a function of rotor angle at 30 RPM.

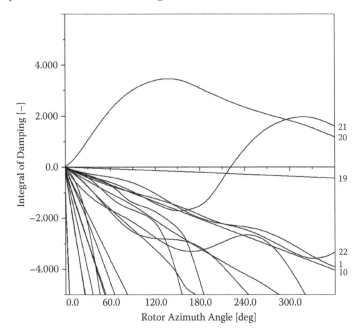

FIGURE 10.84
Integral of the damping at 30 RPM.

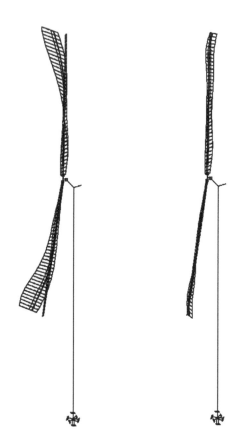

FIGURE 10.85
Real and imaginary parts of the symmetric blade torsion at 30 RPM.

In the two-bladed wind turbines, however, the periodic terms of the coupled system are not cancelled out and the regular complex eigenvalue analysis cannot be used. The stability behavior must be analyzed using the Floquet method discussed in Chapter 5, Section 5.4.

As a test, first let us apply the Floquet method for the uncoupled case. The uncoupled rotor without the tower can of course be calculated by the classical complex eigenvalue method. The real parts of the eigenvalues are shown in Figure 10.86. The solid curves are the complex eigenvalue solutions and have been seen already. The curves with the symbols are obtained by the Floquet method and they are in good agreement.

Let us now consider the coupled model of the two-bladed rotor and the tower. The Floquet matrix described in Chapter 5, Section 5.4 is used to obtain the periodic solution. The canonical transformation described in Chapter 4 Equation (4.26) is applied to the coupled differential equation with time-dependent terms and a system matrix of double size of first order is created. One rotor revolution is divided into k azimuth angles. A delicate compromise must be

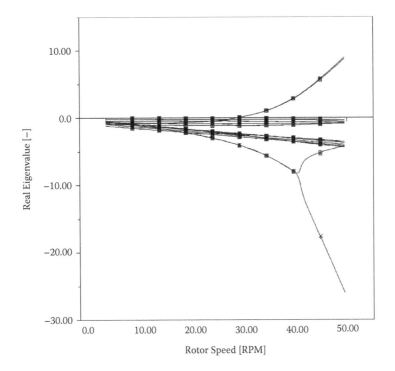

FIGURE 10.86
Comparison of the Floquet method and eigenvalue solution for the rotor only.

struck when selecting k. If k is too small there will be truncation errors in the numerical solution. If k is chosen too large, floating point error accumulation may be detrimental to the result quality.

A transient response analysis is done over these angles of the system with the double-sized system matrix, as shown in Chapter 5, Equation (5.37). The initial condition matrix is the unit matrix. The matrix found after numerical integration is the transition matrix, also called the Floquet matrix.

The eigenvalues of Chapter 5, the Floquet matrix are now calculated with the eigenvalue method of Chapter 5, Section 5.3. The imaginary parts of the eigenvalues are not considered because they are multivalued.

The real parts of the Floquet eigenvalues are called Lyapunov coefficients and they define the stability of the system. The system is unstable when the Lyapunov coefficients are positive. Negative Lyapunov coefficients indicate an unconditionally stable system. When the coefficients are non-negative (≥ 0) then the system is called Lyapunov stable. That means that when the system is perturbed in the neighborhood of an equilibrium point, the solution will remain near the equilibrium point.

The result of the Floquet analysis for the two-bladed wind turbine is shown in Figure 10.87. Clearly, there are three instabilities and they are extracted in Figure 10.88 for better visibility. The symmetric torsion mode (curve 1, 2) is

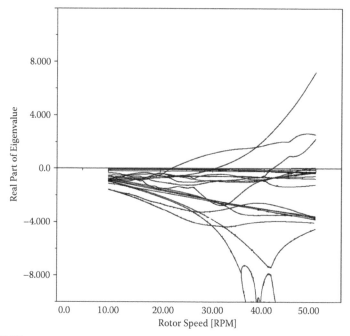

FIGURE 10.87
Real part of the Floquet analysis.

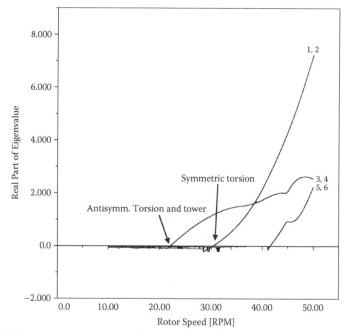

FIGURE 10.88
Real parts of eigenvalues for the three blade instabilities.

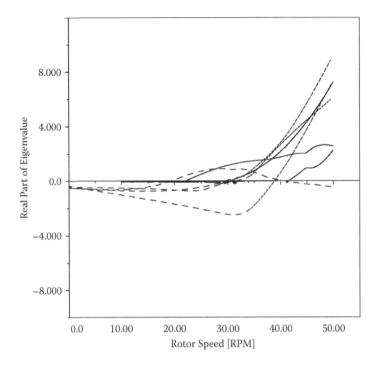

FIGURE 10.89
Real parts of eigenvalues for the three- versus two-bladed rotor.

not influenced by the tower, and the speed of instability is equal to that for the rotor or for the uncoupled blade if the hub and pitch mechanism can be regarded as stiff. The antisymmetric blade torsion (curve 3, 4) is dependent on the tower, and the instability denoted by (5, 6) is also a blade torsion mode, which couples with the tower.

The instabilities found with the Floquet method are slightly higher than those found by the intuitive engineering method of "snap-shot."

The instabilities for the two-bladed turbine can be compared to those of the three-bladed turbine in Figure 10.51. The real parts of the solutions of the three-bladed turbine (dashed lines) and the two-bladed turbine (solid lines) are compared in Figure 10.89. For wind turbines with two blades a strong coupling with the tower occurs, influencing the stability.

There are additional practical scenarios influencing the solution techniques. For example, oblique inflow results in larger periodic loads and turbulence requires stochastic time solutions. These are topics beyond our focus but should be considered by the rotor dynamics engineer.

Appendix

In the development of the transformation matrix introduced in Chapter 1, Equation 1.41 we only used the first order terms of the Taylor series approximation of the sine and the cosine functions. While the $\sin\theta \approx \theta$ approximation is adequate, the cosine term replaced by one has a detrimental effect in some cases as pointed out in Chapter 1, Section 1.7. The following describes the derivation of the transformation matrix with the quadratic term of the cosine approximation retained:

$$\cos\theta = \sqrt{1-\theta^2} \approx 1 - \frac{\theta^2}{2}$$

The differences are demonstrated in Figure A.1.

Applying this, the rotation matrix used in Chapter 1, Equation 1.38 can be written as

$$
\begin{bmatrix}
\cos\theta & -\sin\theta & 0 \\
\sin\theta & \cos\theta & 0 \\
0 & 0 & 1
\end{bmatrix}
\approx
\begin{bmatrix}
1 - \dfrac{\theta^2}{2} & -\theta & \\
\theta & 1 - \dfrac{\theta^2}{2} & \\
& & 1
\end{bmatrix}
$$

The other two rotation matrices are similarly written and the triple product (computed in Chapter 1, Equation 1.41) results in a more complex form of the transformation matrix of Chapter 1, Equation 1.42 as

$$
[\tilde{A}] =
\begin{bmatrix}
0 & -\theta & \psi \\
\theta & 0 & -\varphi \\
-\psi & \varphi & 0
\end{bmatrix}
- \frac{1}{2}
\begin{bmatrix}
\theta^2 + \psi^2 & 0 & 0 \\
0 & \theta^2 + \varphi^2 & 0 \\
0 & 0 & \varphi^2 + \psi^2
\end{bmatrix}
$$

$$
+
\begin{bmatrix}
0 & \varphi\psi & \theta\varphi \\
\varphi\psi & 0 & \theta\psi \\
\theta\varphi & \theta\psi & 0
\end{bmatrix}
= [A_1] + [A_2] + [A_3]
$$

Higher than second order terms occurring during the multiplications were omitted.

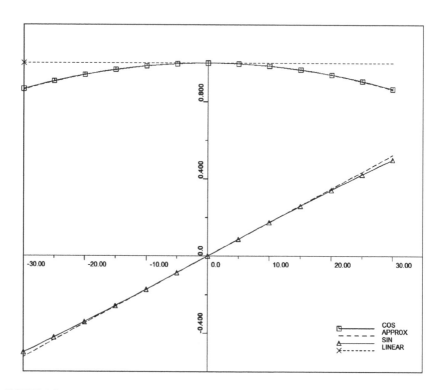

FIGURE A.1
Approximations of sine and cosine functions for small angles

Because this matrix contains quadratic terms, we cannot obtain the transformation matrix in terms of the coordinates of the node's mass points as presented in Chapter 1, Equation 1.45. The triple product contributing to the centrifugal matrix in Equation 1.51 of Chapter 1, Section 1.7 is

$$[\alpha]^T[A]^T[J][A]\{\alpha\}$$

where the matrix is in terms of the mass point coordinates and the vector is in terms of the nodal rotations. In terms of the above quadratic cosine approximation this is of the form

$$\{r'\}^T\left([I]+[A_1]^T+[A_2]^T+[A_3]^T\right)[J]([I]+[A_1]+[A_2]+[A_3])\{r'\}$$

where the transformation matrix components are in terms of the nodal rotations and the vector is in terms of the mass point coordinates.

After executing the multiplication we obtain

$$\{r'\}^T \big([J]+[J][A_1]+[J][A_2]+[J][A_3]$$

$$+[A_1]^T[J]+[A_1]^T[J][A_1]+[A_1]^T[J][A_2]+[A_1]^T[J][A_3]$$

$$+[A_2]^T[J]+[A_2]^T[J][A_1]+[A_2]^T[J][A_2]+[A_2]^T[J][A_3]$$

$$+[A_3]^T[J]+[A_3]^T[J][A_1]+[A_3]^T[J][A_2]+[A_3]^T[J][A_3]\big)\{r'\}$$

By retaining only the linear and quadratic terms while omitting the higher order terms, reordering and grouping we obtain the product as

$$\{r'\}^T \Big([J]+[A_1]^T[J][A_1]+\big([J][A_1]+[A_1]^T[J]\big)+\big([J][A_2]+[A_2]^T[J]\big)$$

$$+\big([J][A_3]+[A_3]^T[J]\big)\Big)\{r'\}$$

The computation of this expression, the evaluation of the derivatives with respect to the nodal rotation and the reversal to present the triple product in terms of the mass point coordinates is rather involved hence not detailed further here. The result ultimately becomes

$$[A]^T[J][A] = \begin{bmatrix} y'^2 - z'^2 & 2x'y' & 2x'z' \\ 2x'y' & x'^2 - z'^2 & 2y'z' \\ 2x'z' & 2y'z' & (x'^2 + y'^2) - (x'^2 + y'^2) \end{bmatrix}$$

This is the accurate form used in Chapter 1, Equation 1.72 to compute the centrifugal matrix.

References

1. Komzsik, L. *Computational Techniques of Finite Element Analysis*, 2nd edition, Taylor & Francis CRC Press, Boca Raton, Florida (2009).
2. Komzsik, L. *The Lanczos Method Evolution and Application*, Society of Industrial and Applied Mathematics, Philadelphia (2003).
3. Gockel, M.A. Practical Solution of Linear Equations with Periodic Coefficients. *Proceedings of Technical Meeting of Western Region of AHS*, Vol. 1, pp. 2–10 (1970).
4. Friedmann, P., Hammond, C.E. Efficient Numerical Treatment of Periodic Systems with Applications to Stability Problems. *International Journal for Numerical Methods in Engineering*, Vol. 11, pp. 1117–1136 (1977).
5. Coleman, R.P., Feingold, A.M. *Theory of Self-Excited Mechanical Oscillations of Helicopter Rotors with Hinged Blades*. NACA Report 1351 (Republished 1957).
6. Gasch, R., Nordman, R., Pfutzner, H. Rotordynamik, 2. *Auflage*, Springer, Berlin (2002) (in German).
7. Someya, T. (Editor). *Journal Bearing Databook*, Springer Verlag, Berlin (1989).
8. Harder, R.L., Desmarais, R.N. Interpolation by Surface Splines. *Journal of Aircraft*, Vol. 9, No. 2, pp. 189–191 (1972).
9. Komzsik, L. *Approximation Techniques for Engineers*, Taylor & Francis CRC Press, Boca Raton, Florida (2007).
10. Bisplinghoff, L.R., Ashley, H., Halfman, R.L. *Aeroelasticity*, Addison-Wesley Publishing Company, Reading, Massachusetts (1957).
11. Rodden, W.P. *Theoretical and Computational Aeroelasticity*, Crest Publishing, Cararillo, California (2011).
12. Fung, Y.C. *Aeroelasticity*, Dover Publications, New York (1969).
13. Scanlan, R.H., Rosenbaum, R. *Aircraft Vibration and Flutter*, Dover Publications, New York (1968).
14. Dowell, E.H., et al. *A Modern Course in Aeroelasticity*, Kluwer Academic Publishers, Dordrecht (2004).
15. Boeswald, M., Vollan, A., Govers, Y., Frei, P. *Solar Impulse—Ground Vibration Testing and Finite Element Model Validation of a Lightweight Aircraft*. IFASD-2011-132, (2011).
16. Theodorsen, T. *General Theory of Aerodynamic Instability and the Mechanism of Flutter*. NACA Report 496 (1935).
17. Küssner, H.G., Schwarz, L. Der Schwingende Flügel mit Aerodynamisch Ausgeglichenem Ruder. *Luftfaht-Forschung*, Band 17 (Dezember 1940).
18. Hassig, H.J. An Approximate Time Damping Solution of the Flutter Equation by Determinant Iterations. *Journal of Aircraft*, Vol. 8, No. 11, pp. 555–559 (November 1971).
19. Glauert, H. Airplane Propellers, Division L. in: Durand, W.F. *Aerodynamic Theory*, 1935. Republished by Dover Publications, New York (1963).
20. Arendsen, P. Final Report of the GARTEur Action Group (SM) AG-21 on "Multidisciplinary Wing Optimisation," NLR-TR-2001-557 NLR, Amsterdam (2001).

21. Schneider, G., van Dalen, F., Krammer, J., Stettner, M. *Multidisciplinary Wing Design of a Regional Aircraft Regarding Aeroelastic Constraints*. Paper presented at the Optimization in Industry Conference-II, Banff (Canada) (6-11 June 1999).
22. Vollan, A. Aeroelastic Analysis of a Transport Aircraft. *Luftfahrttechnisches Handbuch LTH*, BM 45 100-1 (2010).
23. Ribner, H.S. *Propellers in Yaw*. NACA Report 820 (1945).
24. Houbolt, J.C., Reed, W.H. Propeller-Nacelle Whirl Flutter. *Journal of the Aerospace Science*, pp. 333–346 (March 1962).
25. Reed, W.H., Bland, S.R. *An Analytical Treatment of Aircraft Propeller Precision Instability*. NASA TN D-65 (January 1961).
26. Rodden, W.P., Rose, T.L. *Propeller/Nacelle Whirl Flutter Addition to MSC/Nastran*, The MacNeal-Schwendler Corporation, Los Angeles (1978).
27. Bland, S.R., Bennett, R.M. *Wind Tunnel Measurement of Propeller Whirl-Flutter Speeds and Static-Stability Derivatives and Comparison with Theory*. NASA TN D1807 (August 1963).
28. Hau, E. *Wind Turbines*, 2nd edition, Springer Verlag, Berlin (2006).
29. Wilson, R.E., Lissamann, P.B.S., Walker, S.N. *Aerodynamic Performance of Wind Turbines*. ERDA NSF 04014-76 1. U.S. Depatment of Energy (1976).
30. Lobitz, D.W. Aeroelastic stability predictions for a MW-sized blade. *Wind Energy*, Vol. 7, pp. 211–224 (September 2004).
31. Pedersen, P.T. On Forward and Backward Precession of Rotors. *Ingenieur-Archive*, Vol. 42, pp. 26–41 (1972).
32. Hutin, P.-M. *Application de la Méthode "Modes Partiels" a la Prévision du Comportement Vibratoire d`un Hélicoptère en Vol Communication Présentée au Groupe de Travail de l`AGARD sur les Structures des Hélicoptères*, Tönsberg, Norvège (1970).
33. Genta, G. *Dynamics of Rotating Systems*, Springer, New York (2005).

Index